SpringerBriefs in Environmental Science

SpringerBriefs in Environmental Science present concise summaries of cutting-edge research and practical applications across a wide spectrum of environmental fields, with fast turnaround time to publication. Featuring compact volumes of 50 to 125 pages, the series covers a range of content from professional to academic. Monographs of new material are considered for the SpringerBriefs in Environmental Science series.

Typical topics might include: a timely report of state-of-the-art analytical techniques, a bridge between new research results, as published in journal articles and a contextual literature review, a snapshot of a hot or emerging topic, an in-depth case study or technical example, a presentation of core concepts that students must understand in order to make independent contributions, best practices or protocols to be followed, a series of short case studies/debates highlighting a specific angle.

SpringerBriefs in Environmental Science allow authors to present their ideas and readers to absorb them with minimal time investment. Both solicited and unsolicited manuscripts are considered for publication.

Azimah Ismail · Hafizan Juahir

Advanced Chemometrics

Unlocking Patterns in Oil Spills

Azimah Ismail
Department of Manufacturing Engineering Technology
Faculty of Innovative Design and Technology
Universiti Sultan Zainal Abidin
Kuala Nerus, Terengganu, Malaysia

Hafizan Juahir
Faculty of Bioresources and Food Industry
Universiti Sultan Zainal Abidin
Besut, Terengganu, Malaysia

ISSN 2191-5547 ISSN 2191-5555 (electronic)
SpringerBriefs in Environmental Science
ISBN 978-3-032-10741-1 ISBN 978-3-032-10742-8 (eBook)
https://doi.org/10.1007/978-3-032-10742-8

This Springer imprint is published by the registered company Springer Nature Switzerland AG
The registered company address is: Gewerbestrasse 11, 6330 Cham, Switzerland

This book is dedicated to our families, colleagues, and students, whose unwavering support and curiosity have fueled our pursuit of knowledge.

To the research community of Universiti Sultan Zainal Abidin, the East Coast Environmental Research Institute, and all collaborators who have shared their expertise and passion for environmental stewardship.

And to communities around the scientists, policymakers, educators, and citizens, who work tirelessly to protect our waters, restore ecosystems, and safeguard the planet for future generations.

May this work inspire a united effort across borders, harnessing the power of science, innovation, and integrity to solve complex environmental mysteries—unlocking the truth, one dataset at a time.

Acknowledgements We extend our deepest gratitude to the Universiti Sultan Zainal Abidin (UniSZA), Faculty of Innovative Design and Technology, the East Coast Environmental Research Institute (ESERI), Faculty of Bioresources and Food Industry, for their unwavering institutional support. Special thanks go to our colleagues, whose dedication and insights have enriched this work. We are grateful for the research grants from Integrated Envirotech Sdn. Bhd. as a financial funder and the Chemistry Department of Malaysia and the Department of Environment, Malaysia, whose support made this project possible both within Malaysia and internationally, shared their expertise and data. Lastly, we would like to thank our families for their patience, encouragement, and understanding throughout the long process of writing and compiling this book. Without their support, this book would not have come to fruition.

Competing Interests The authors declare there are no competing financial or personal interests that could have influenced the work presented in this book. All analyses, interpretations, and conclusions are based solely on the scientific evidence from the real case studies of oil spills and data available to the authors.

Ethics Approval This work involves the analysis of environmental datasets collected in accordance with local and international research ethics guidelines. Where applicable and permissions were obtained from relevant authorities for the use of data. No human participants or live animals were involved in the studies presented in this book.

About This Book

Oil spills present one of the most pressing environmental challenges of our time, threatening marine biodiversity, coastal economies, and human well-being. Investigating these incidents requires precision, scientific rigor, and the ability to extract meaningful patterns from complex and multidimensional datasets. This book introduces readers to the decisive role of chemometrics in environmental forensics. It also bridges environmental science and advanced data analytics to equip readers with both the theoretical foundations and the practical skills needed to unravel oil spill cases. Drawing upon years of research and real-world case studies, the authors Azimah Ismail and Hafizan Juahir demonstrate how multivariate statistical analysis, pattern recognition, and predictive modeling can be applied to identify pollution sources, characterize spill composition, and support legal and policy decision-making. While oil spill forensics is the primary focus, the methods and principles discussed extend to a wide range of environmental applications, from water quality monitoring to industrial process optimization. Therefore, comprehensive coverage of chemometric techniques from basic to advanced levels has been introduced and served as practical guidance for researchers, practitioners, and policymakers in applying these techniques to environmental problems.

Contents

About the Authors

Ir. Dr. Azimah Ismail is a Senior Lecturer at the Faculty of Innovative Design and Technology, Universiti Sultan Zainal Abidin (UniSZA), Gong Badak Campus, Terengganu. Previously, she served as the Academic and Postgraduate Coordinator at the East Coast Environmental Research Institute (ESERI). Her areas of specialization span Activated Carbon, Organic Composites, Oil Spill Fingerprinting, Data Analytics, Chemometrics, Six Sigma, and Water and Wastewater Treatment Technology, with strong expertise in Material Science, Heavy Metals, Environmental Chemistry, and Chemical and Process Engineering. She is also experienced in chemical process optimization and wastewater process of engineering applications.

Her research extends to applied science and technology in environmental fields, particularly in developing innovative solutions for pollution control, resource recovery, and sustainable manufacturing. She is a member of the Special Interest Group under the Cluster of Material Science and Nanomaterials at UniSZA, contributing to multidisciplinary research at the intersection of materials, environment, and process engineering.

Professor Dr. Hafizan Juahir is a Professor at the Faculty of Bioresources and Food Industry, Universiti Sultan Zainal Abidin (UniSZA), Terengganu, and a leading researcher in environmental modeling, water quality assessment, and chemometrics. His work focuses on applying advanced statistical, computational, and geospatial techniques to monitor, analyze, and predict environmental quality at local, regional, and national scales. Specializing in water quality modeling, environmental data analytics, multivariate statistical analysis, and spatial-temporal modeling, he has contributed extensively to environmental management through the development of decision-support systems and predictive models for pollution control. His research integrates GIS, remote sensing, and statistical modeling to address pressing environmental issues, particularly in water resources, air quality, and ecological risk assessment.

He is also recognized for his expertise in chemometrics applications to environmental datasets, enabling more accurate interpretation of complex environmental processes. His innovative approaches support policymakers, industries, and researchers in achieving sustainable environmental management. He has been actively involved in national and international collaborations, producing numerous high-impact publications and supervising postgraduate research in environmental science, data analytics, and sustainable resource management.

Chapter 1
Recognition Using Numerical Methods Approach

Abstract This chapter described the general explanations of the recognition process using numerical methods and elaborated on the factors and impact of oil spill pollution into environment. This is an essential point to brief the readers that oil spill is the most severe impact, especially causing widespread damage to ecosystems, marine biodiversity, human health and global economy. Mostly, this accident happens when petroleum or its by products are unintentionally or deliberately released into the marine or coastal environment. This chapter insights contribute significantly to environmental protection and for sustainable marine resource management. Polyaromatic hydrocarbon (PAH) is one of the chemical organic compounds that are formed by fused benzene rings without heteroatoms or substituents. PAH is also being brief in this chapter as a critical indicator of environmental pollution. Their toxicity, persistence and diagnostic value in forensic investigation make them a central focus in both analytical chemistry and environmental management. Their behavior and fate are also essential for impact assessment, regulatory compliance, and long-term ecosystem protection.

Keywords Polyaromatic hydrocarbons · Marine · Heteroatoms · Biodiversity · Environmental forensic

1.1 Overview of Oil Spill Contamination

One of the most destructive types of environmental pollution is oil spill contamination, which has a profound effect on ecosystems, marine life, human health, and economy. These events take place when petroleum or its by products are unintentionally or purposely discharged into the environment, mainly into coastal and marine environments. Developing successful mitigation and remediation techniques requires an understanding of the causes, traits, and effects of oil spill pollution.

Numerous human activities can result in oil spills, including:

- Accidental spills from oil tankers during transit, crashes, or groundings are all considered maritime transport.

A. Ismail and H. Juahir, *Advanced Chemometrics*,
SpringerBriefs in Environmental Science, https://doi.org/10.1007/978-3-032-10742-8_1

Fig. 1.1 Oil spillage on the beach. *Source* www.google.com/ November 28, 2024, 1.42 PM

- Equipment malfunctions or blowouts during offshore drilling operations might occur when extracting oil from sea beds.
- Pipeline, storage tank, or refinery leaks are examples of industrial activities.
- Accidents and Natural Disasters: Inadvertent spills may result from ship mishaps, hurricanes, or tsunamis.
- Intentional Releases: military conflicts, unlawful dumping, or sabotage.

The effects of oil spills on beaches (Fig. 1.1) provide a serious threat to environmental contamination and contamination on a large scale. Additionally, it can kill birds and coat rocks, sands, and plants. Long-term exposure to it can also be hazardous to humans, endangering their health.

1.2 Polyaromatic Hydrocarbons (PAHs) in Oil Spills

A class of chemical molecules made up of many aromatic rings which known as polyaromatic hydrocarbons, or PAHs. They are a key problem in the context of oil spills because they are a large hazardous component of crude oil and petroleum products (Kappell et al. 2014). The environment, marine life, and human health are all seriously endangered by PAHs, which are persistent and bioaccumulative. The structure and properties of PAHs, which are made up of fused benzene rings, vary from two-ring compounds like naphthalene to larger, more intricate molecules like benzo[a]pyrene. They prefer to stick to sediments or organic debris because they are hydrophobic, which means they do not dissolve easily in water.

1.3 Chemical Structure and Classification of PAHs

In common, PAH categorization uses ring numbers as follows:

- PAHs of low molecular weights (LMWs)

Comprise of two or three rings, such as phenanthrene, anthracene, or naphthalene. It contains more water-soluble and volatile than bigger PAHs and less harmful than HMW PAHs.

HMWs, or high molecular weight PAHs, consists of four or more rings, such as pyrene, chrysene, or benzo[a]pyrene. More persistent, poisonous, and bioaccumulation-prone. Moreover, it is less hydrophobic and volatile which commonly found in thicker residues and oils. Figure 1.2 Molecular structure of benzo[a]pyrene, a high molecular weight polycyclic aromatic hydrocarbon (PAH). This compound is a key marker for petrogenic and pyrogenic pollution sources and is often monitored due to its toxicity and environmental persistence.

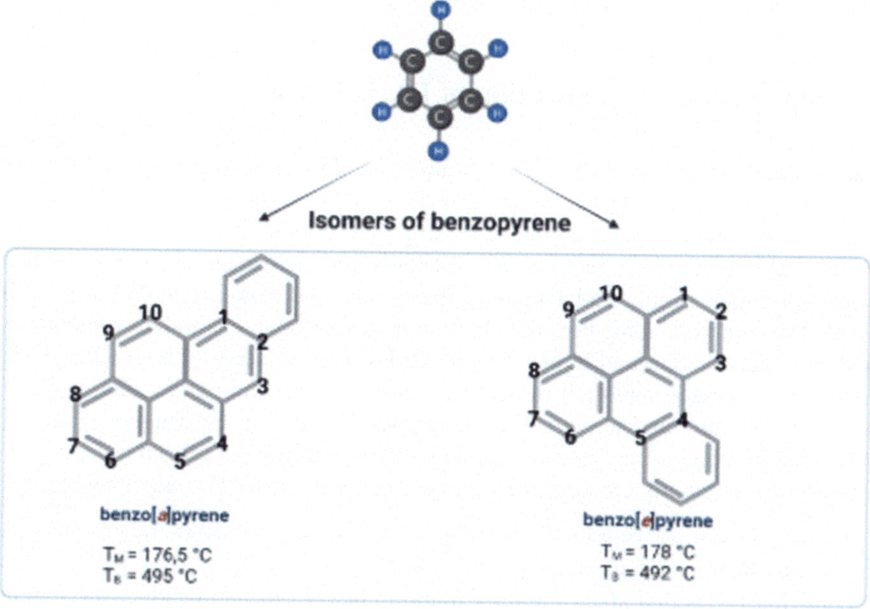

Fig. 1.2 High molecular weight PAHs (benzo[a]pyrene). *Source* www.google.com, November 29, 2024/ 10.48 PM

Fig. 1.3 Oil Spill Event in Malaysia. *Source* www.google.com/ November 28, 2024, 1.55 PM

Figure 1.3 illustrates the environmental impact of a local oil spill incident, emphasizing the importance of monitoring, rapid response, and remediation measures to protect coastal ecosystems.

1.4 Sources and Composition of PAHs in Oil

Crude oil and petroleum derivatives naturally contain a complex group of chemical molecules called polyaromatic hydrocarbons, or PAHs. They may also develop as a result of incomplete combustion. Assessing their effects on the environment and human health, particularly in relation to oil spills, requires an understanding of their composition and origin. Understanding the origins and makeup of PAHs in oil is essential to comprehending how they behave in the environment and the effects they have on the ecosystem and human health. PAHs are produced by both natural and man-made processes, and their molecular weight and structure affect their toxicity, environmental fate, and remediation techniques. To reduce the dangers related to PAH contamination, effective management and monitoring are crucial.

Crude oil frequently contains the highly volatile two-ring PAH naphthalene.

- Two common three-ring PAHs found in crude oil and combustion by products are phenolphthalein and anthracene.
- Four-ring PAHs that are commonly found in heavy oils and residues are pyrene and chrysene.
- One five-ring PAH that is known to cause cancer is benzo[a]pyrene.

1.5 Natural and Anthropogenic Sources of PAHs

LMW predominates among petrogenic PAHs, which are more easily dissolved or volatilized and have transient effects in aquatic environments. Enriched with HMW PAHs, pyrogenic PAHs are toxic and persistent, causing long-term environmental harm as well as health issues. Mixed sources of petrogenic and pyrogenic PAHs are frequently present in oil spills, making it more difficult to identify the source and evaluate the danger. They come from both man-made and natural sources.

- Natural sources, such as forest fires and volcanic eruptions, are usually localized and intermittent.
- Anthropogenic sources are more pervasive, ongoing, and frequently found in industrial and urban areas.

1.5.1 Sources from Nature Volcanic Activity

PAHs are released into forest fires when organic components in the ground burn during volcanic eruptions. A forest fire, sometimes referred to as a wildfire, is an uncontrolled fire that spreads quickly through grasslands, forests, and other natural settings. Natural occurrences may be the cause of these fires such as volcanic eruption and spontaneous combustion due to lightning. The generation and release of PAHs into the environment are influenced by the natural changes that occur in buried sediments throughout time or scientific name is called diagenesis. Diagenesis is the process by which PAHs are produced when plants and organic debris are not completely burned during forest fires. When organic matter decomposes in sediments under pressure and heat over geological timescales, PAHs may naturally form.

1.5.2 Anthropogenic Sources

The primary source of polycyclic aromatic hydrocarbons (PAHs) in the environment is human activity, which involves the burning or processing of organic resources. Because these man-made sources contribute significantly to environmental contamination and have consequences for human health, they have been the subject of current studies. Below describes the significant environment:

1.5.2.1 Fuel Combustion from Fossils

In both urban and industrial environments, burning fossil fuels including coal, oil, and natural gas are a significant producer of PAHs. When coal and other carbon-based fuels are burned in thermal power plants and industrial boilers, PAHs are released.

Large volumes of PAHs are released into the atmosphere when coal and wood are used for home heating (domestic heating).

1.5.2.2 Transportation Industry

PAHs are released through exhaust systems when gasoline and diesel fuels in automobile engines are not completely burned. Higher PAH emissions are specifically associated with diesel engines. Likewise, the burning of aviation fuels and heavy oils in ships contributes to the release of PAHs.

1.5.2.3 Industrial Processes

In the steel production and coke process, coal is heated to high temperatures, which releases PAHs into the atmosphere. Petrochemicals and oil refining are the two major sources of these harmful pollutants during the production of petrochemicals and the processing of crude oil. Another industry attributable is that asphalt production. By heating bituminous materials, asphalt is produced and used in road construction, releasing PAHs.

1.5.2.4 Burning Waste and Biomass

Burning garbage and organic materials is common in many areas and significantly increases the emissions of PAHs. Agricultural residue burning are also released the toxin into the atmosphere when crop residues are burned after harvest. The incineration of municipal and industrial waste, especially when the combustion conditions are not ideal, is also subject to the toxin pollutants. Likewise, the residential biomass burning. Homes that use wood, charcoal, and other biomass fuels for heating and cooking produce a lot of PAHs.

1.6 Analytical Techniques for PAH Detection and Quantification

A class of chemical molecules known as polycyclic aromatic hydrocarbons (PAHs) poses serious risks to human health and the environment because of their mutagenic and carcinogenic qualities. Understanding the distribution, sources, and possible hazards of PAHs depends on the precise detection and quantification of these chemicals in a variety of matrices, including air, water, soil, sediments, and biological tissues. The general advanced analytical methods that are complimentarily used in this field are listed below:

1.6.1 Chromatographic Methods

The most effective method for identifying and evaluating individual PAHs in complicated mixtures is chromatography. Capillary columns are used in gas chromatography (GC) to achieve high resolution (Park et al. 2023). It is frequently used in conjunction with detectors such as mass spectrometry (MS) or flame ionization detectors (FIDs). Both volatile and semi-volatile PAHs can use it.

1.6.2 Coupled Methods of Mass Spectrometry

Using mass spectrometry (Fig. 1.4) in conjunction with chromatography improves selectivity and sensitivity:

By using their mass-to-charge ratio, GC–MS makes it possible to identify and quantify PAHs. Because of its strong performance, environmental samples are frequently employed. As analytical methods for PAH detection and quantification advance, they offer increased sensitivity, precision, and dependability. The procedure selected is determined by the intended use, PAH concentration, and sample matrix. Developments in spectroscopic, sensor-based, and chromatographic technology promise more advancements in PAH investigations connected to the environment and human health. PAH detection and quantification analytical methods are constantly improving in terms of sensitivity, accuracy, and dependability. The targeted application, PAH concentration, and sample matrix all influence the procedure selection.

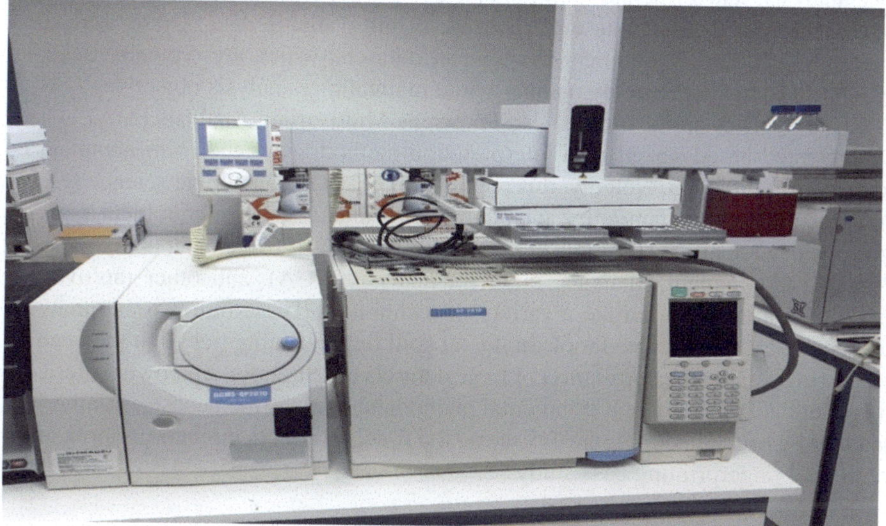

Fig. 1.4 Gas chromatography-mass spectrometry. www.google.com/ December 5, 2024, 6.24 P.M.

Technological developments in chromatography, spectroscopy, and sensors promise more advancements in PAH investigations connected to the environment and human health.

1.6.3 Data Interpretation and Quality Control

Deuterated PAHs are frequently employed as internal standards in order to take into consideration variations in extraction and analysis. In order to ensure accuracy, method validation takes into account characteristics like precision, limit of quantification (LOQ), recovery, and limit of detection (LOD).

Multivariate Data Analysis: Principal Component Analysis (PCA) is one tool that helps comprehend complex datasets and pinpoint the origins of PAHs.

1.7 Advances in Fingerprinting Techniques for Oil Spill Analysis

Chemometrics has gained widespread recognition for its powerful statistical approaches and mathematical procedures. The application is not only limited in chemistry, but in all fields as well such as finance and engineering. The example applications in particular are for: (a) classification or clustering of the chemical compound homogeneity and non-homogeneity; (b) pattern recognition, discriminatory analysis, and Principal Component; and (c) reactant and reaction modelling and monitoring that permits extensive use of all available data. These were used to resolve and show complicated chemical datasets, which are typically unstable dynamic systems, in a timely manner while maintaining analysis objectives.

Chemometrics application is also known as Multivariate Analysis (MA), which implemented an advanced data interpretation method generated from abundant or complex mixture data. Notedly, this application offers the analytical science resolution processes, both quantitative and qualitative evaluation. The hierarchical agglomerative cluster analysis (HACA), Discriminant Analysis (DA), cluster analysis (CA), Principal Component Analysis, or factor analysis (PCA), and other multivariate modelling approaches utilize a variety of techniques. These multivariate analyses are such more sophisticated tools in the oil spill fingerprinting field that they can be used to investigate the correlation of large datasets to search for unknown patterns of chemical data related to environmental phenomena such as the genesis, weathering, and mixing of crude oil. Figure 1.5 shows the flowchart of chemometrics application in oil spill fingerprinting case study.

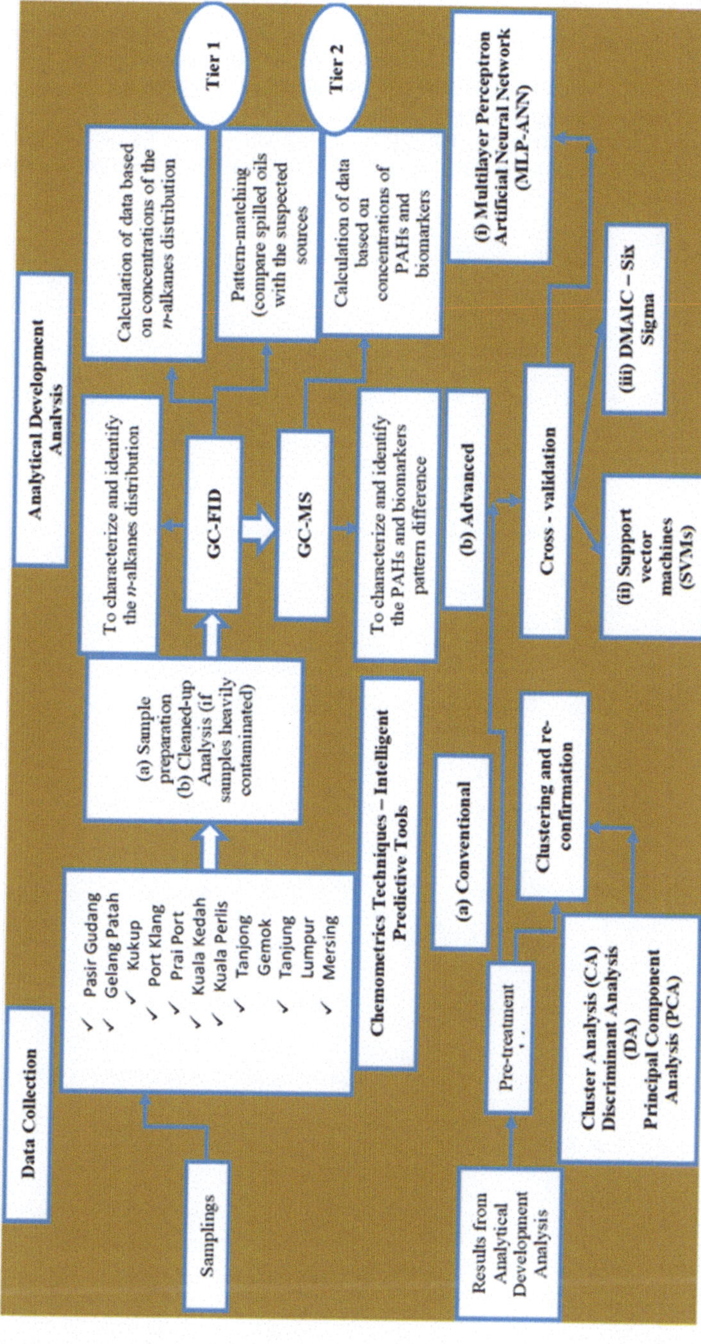

Fig. 1.5 Flowchart of chemometrics application in oil spill fingerprinting case study. *Source* Authors

References

Kappell AD, Wei Y, Newton RJ, Van Nostrand JD, Zhou J, McLellan SL, Henson MW, Moran MA, Kostka JE (2014) The polycyclic aromatic hydrocarbon degradation potential of Gulf of Mexico native coastal microbial communities after the Deepwater Horizon oil spill. Front Microbiol 5:205. https://doi.org/10.3389/fmicb.2014.00205

Park J, Kim K, Ryu D, Whang JH, Mah JH (2023) GC-MS/MS method for determination of polycyclic aromatic hydrocarbons in herbal medicines. Molecules 28:93853. https://doi.org/10.3390/molecules28093853

Chapter 2
Classification of Similar Chemical Properties in Oil Spills

Abstract This chapter explains the practically how to classify the oil spill data which identifies the similarities in the chemical properties using chemometric analysis. This chapter aimed to make a detailed understanding of the source and also behavior of spilled oil based on its chemical composition. Last but not least, this chapter emphasizes that accurate classification of oil spills requires rigorous data pre-processing to ensure high-quality analysis. Standardizing variables, handling outliers, and confirming normality are critical steps in producing meaningful and reproducible clustering results for environmental forensic investigation. The classification of oil spill dataset is imperative to determine the similarities of chemical properties within the cluster or dissimilarities with different clusters. It can be done using the chemometric analysis. However, prior to proceeding with the classification analysis, pre-treatment is another approach that must be successfully performed to enable the data properly center, scale, and reduce the dimensionality of the data. By doing this, it can improve predictive modeling and data quality, anomalies during the pre-treatment phase of oil spill data, while missing values are replaced with the mean (for continuous data) and median (for skewed data) techniques. Skewed data refers to the distribution of oil spill datasets if it is not symmetrical around the mean. For instance, the histogram or probability distribution has an uneven shape because it is "pulled" or "stretched" more to one side. Standardized skewness and standardized kurtosis are calculated for data verification that the data distribution is normal. Skewness quantifies how asymmetrical the data distribution is. It shows if the data points are distributed more to the left or right of the mean. Kurtosis provides insights about the shape of frequency distribution, data distribution's "tailedness." It aids in comprehending the existence of outliers. In fact, both skewness and kurtosis as additional variables to see how they influence clustering. The data or variables' goodness of fit to a normal distribution is assessed using the 95% confidence level normality tests, namely the Shapiro–Wilk, Anderson–Darling, Liliforths, and Jarque–Bera tests. The results of the tests will normally show that not all of the raw data is normally distributed ($p < 0.05$). The data is optimally transformed using log-transformation and z-scale normalized (mean = 0, standard deviation = 1). Standardization is crucial because the numerical data or parameters are subject to

A. Ismail and H. Juahir, *Advanced Chemometrics*,
SpringerBriefs in Environmental Science, https://doi.org/10.1007/978-3-032-10742-8_2

different unit ranges and measurements. Additionally, it makes it possible to minimize the impact of measurement discrepancies and guarantee that every variable has an equal effect. Data pre-treatment is essential because it guarantees that data is clear, consistent, and prepared for insightful analysis.

Keywords Predictive modeling · Pre-treatment · Kurtosis · Chemometric · Skewness

2.1 Cleaning of Data

Eliminate missing data: Determine if there are any missing values and choose a method. It can be performed by eliminating rows or columns that include missing data, replacing them with mean or median values, or applying correct strategies like k-nearest neighbors.

Deal with Outliers: Results can be distorted by outliers, particularly in small datasets. Determine whether to retain, alter, or eliminate outliers based on their influence by using methods like the z-score or IQR (interquartile range).

Correct Mistakes: Particularly in categorical variables (e.g., "Oil Type" vs. "Oil Type" vs. "O"), look for errors or inconsistent data entries.

Data Transformation Normalization/Standardization: Apply scaling techniques to features with different ranges:

- Rescaling values to fall within a range, often [0, 1], is known as normalization.
- Standardization: Assigns a mean of 0 and a standard deviation of 1.0 to the data.
- Categorical variable encoding: Employ strategies such as ordinal encoding for ordered categories or one-hot encoding for nominal variables.
- Log Conversion: Applying a log (or square root) transformation can normalize distributions for data with skewed distributions, particularly in small datasets.
- Check for Correction: Particularly in categorical variables (e.g., "Oil Type" vs. "Oil Type" vs. "O"), look for errors or inconsistent data entries.

Equation 2.1 represents the Z-scale transformation (Kannel et al. 2007):

$$x = \frac{x - x'}{\sigma} \tag{2.1}$$

where Z is the standardized value, *x'* is the variable's average value, σ is the standard deviation, and *x* is the original value of a measured parameter.

The step by step to check the missing data can refer to the following:

1. Firstly, open the Excel workbook and insert the oil spill dataset

   Click Find & Select.

2. Click Find & Select

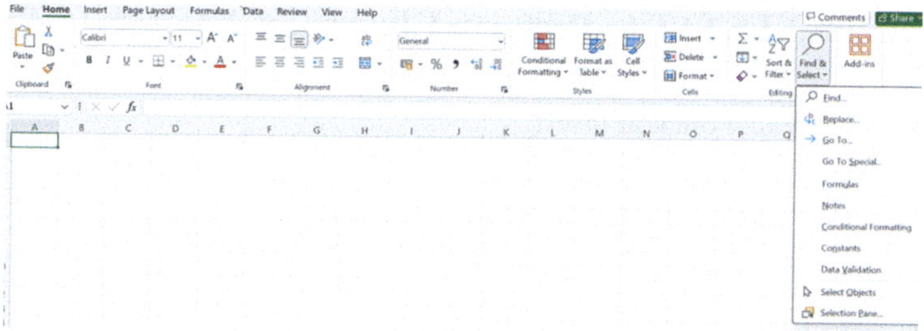

3. Next click Replace

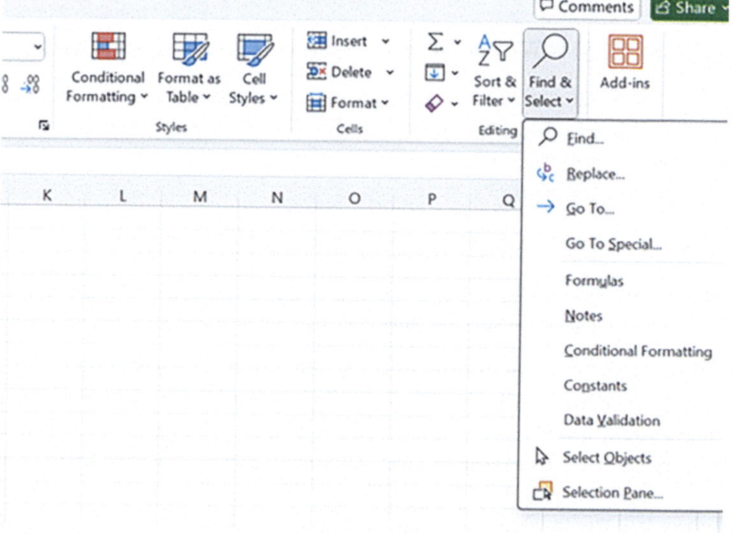

Find the odd values from the raw dataset such as < 0.005 and then replace it with (0.005/2) + 1 for data normalization (Templ. 2023).

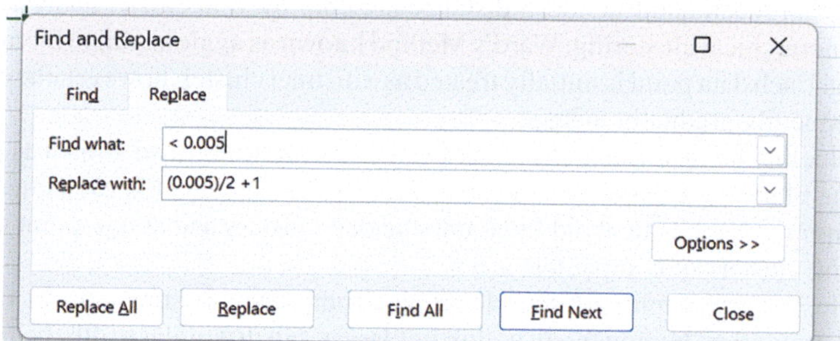

2.2 Launch XLSTAT to Perform Clustering Analysis (CA)

1. Firstly, go to Excel click at insert

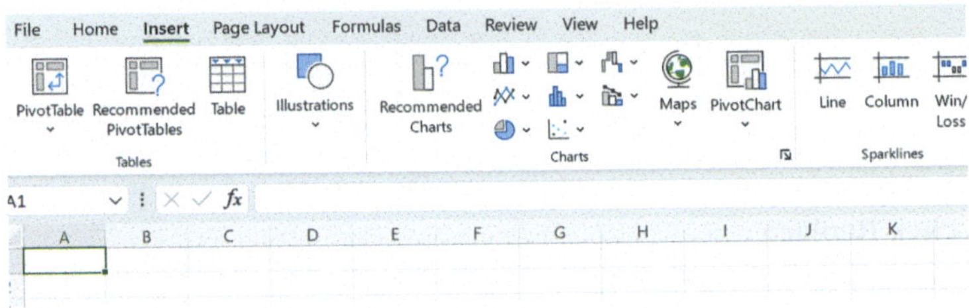

2. Then, use XLSTAT's Ward's Method by clicking Add-in launch the XLSTAT.
   - Get oil spill dataset ready. Make sure the data is clear and well-structured, usually with variables in columns and observations in rows.
   - Open Excel and select the XLSTAT tab.
   - Then, go to Clustering.
     - XLSTAT → Data Analysis → Clustering Hierarchically
   - Choose the Information
   - Select the range of data to be analyzed.
     - If headers are present in your data, select the Labels option.
     - Configure the Clustering Settings. The criterion for aggregation is Ward's approach.
     - Measure of Dissimilarity, by selecting a suitable distance metric, the Euclidean distance.
     - Options for Output. Indicate the location of the dendrogram and cluster analysis report.
     - Execute the analysis. To start the clustering process, click OK.

Ward's approach and Euclidean distance are employed in cluster analysis, especially in hierarchical clustering. Ward's Method known as agglomerative hierarchical clustering. Each data point is initially treated as a distinct cluster; however, the clusters are subsequently gradually merged.

By reducing the increase in the sum of squared deviations (also known as inertia) at the time of cluster merger, it seeks to reduce the variance within each cluster.

This approach tends to avoid huge imbalanced clusters and single-point clusters by forming clusters of comparable sizes.

The clustering/grouping of the oil samples/compounds is subject to its intrinsic chemical properties' homogeneity within its classes, but dissimilar to other clusters or groups. Cluster analysis (CA) is also known as the unsupervised pattern recognition method, which enables reliable classification of the oil spill samples into clusters.

The oil spill samples on the normalized datasets are clustered using the hierarchical agglomerative cluster analysis (HACA). By integrating clusters that offer the fewest additional "total within-cluster classification points according to their shortest Euclidean distances," Ward's Method and Euclidean distances work together to minimize the total within-cluster variance. The process happens through merging clusters iteratively until there is only one cluster left or a halting requirement is satisfied while minimizing the variance increase.

- Calculating the Euclidean distances between each pair of data points.
- Grouping points into clusters based on the smallest Euclidean distances.
- Iteratively merging clusters in a way that minimizes the increase in variance until only one cluster remains or a stopping criterion is met.

The clustering or grouping variables in the oil spill case are performed based on its similarity of properties or having similar-looking fingerprints found. The oil samples with similar chemical properties and dissimilar with others are clustered together that is demonstrated in the form of a dendrogram as in Fig. 2.1.

The observation of the clusters referring to the oil spills case was perfectly elaborated as in Fig. 2.2. The dendrogram generated by cluster analysis (CA) shows four spatial clusters due to the spatial patterns of oil spill observed at each sampling stations along the Straits of Malacca. However, Clusters 1 and 3 show very high similarity in chemical properties, and they could possibly be combined into one group. Evidently, the quantity of several waste oils (MO, Bunker C, and diesel-based

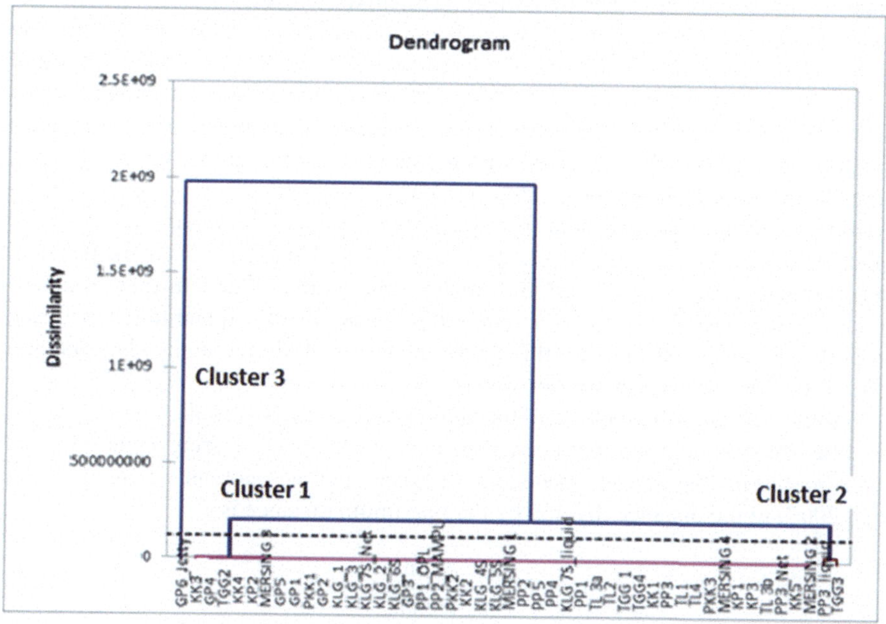

Fig. 2.1 Clustering of the oil samples by group according to its chemical properties

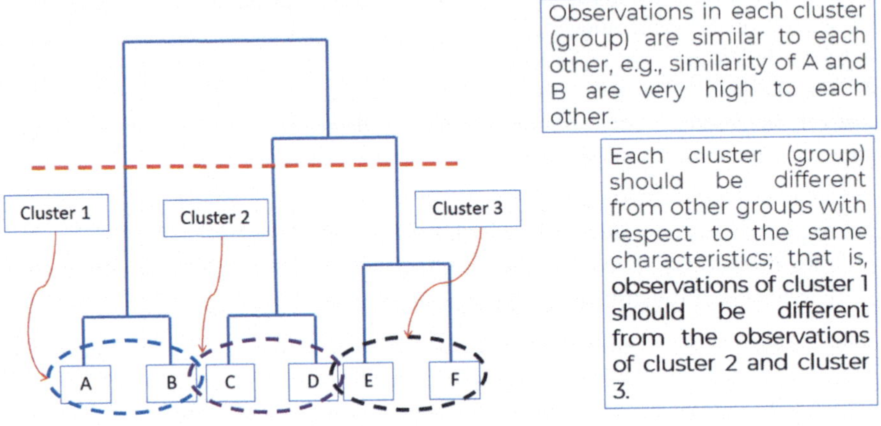

Fig. 2.2 Observations in each cluster

products) dominated Cluster 1. Both the distinctive fingerprint in the CA spill source classification and the diversity of waste oils (MO, Bunker C, lubricating oil, and diesel-based products) were found to be prevalent in the sample of oil leaked sources (Asif et al. 2022).

According to the above profile plot, these greasy wastes contained the lowest quantities of alkane constituents. Lubricating oil mixed with diesel-based items dominated Cluster 2, while the diesel-type product was displayed in Cluster 3. The profile plots of Clusters 2 and 3 showed a fingerprint that was almost identical but had distinct features in the fragmentation of hydrocarbon-fuel alkanes. It also illustrates the significant variations in the petroleum hydrocarbon oil contents of the n-alkanes compounds were closely connected with the classification of diesel-based products from Clusters 3 (GP_Jetty) and 2 (PP3-liquid). These profiles showed likely characteristics of oil that was separated from fresh diesel oil and the weathered diesel oil effect, which is caused by prolonged exposure and has led to changes in the physical and chemical properties of the oil. Before the combination is done, further investigation is needed. Visual inspection is required to ensure that the quantitative evidence supports each decision made in this study. This case study recommended line and boxplot graphs as visual inspection tools.

Finding variables that belong to the same group and have similar characteristics but differ from other groups has been made possible by this clustering technique. The y-axis represents the squared Euclidean distance, and the quotient linkage (Omran et al. 2007) to the distance divided by the maximum distance is;

$$\text{Euclidean} = \frac{D\,\text{link}}{D\,\text{max}} \times 1000 \tag{2.2}$$

The Euclidean distance in chemometrics is a measurement of the linear separation between two points in a multidimensional space, which are usually samples, variables, or observations. Quantifying the similarity or dissimilarity of data points based on their numerical values is a popular application. The Euclidean distance, d, between two points is $x = (x_1, x_2, \ldots, x_n)$ and $d = (d_1, d_2, \ldots, d_n)$. It is given that $y = (y_1, y_2, \ldots, y_n)$ in an n-dimensional space:

$$d(x, y)\sqrt{\sum_{i=1}^{n}(x_i - y_i)^2} \tag{2.3}$$

Essential Elements of Chemometric Data Representation: Points are frequently vectors that reflect chemical measures in chemometrics, such as chromatographic profiles, concentrations, or spectrum data. Its applications are many in clustering algorithms that is putting together examples that are similar.

However, for the analysis of Principal Components figuring out distances and scores in smaller dimensions. Finding samples that are distant from the primary data cluster is known as outlier identification.

Normalization: Because variables may have different units or ranges, which might significantly affect the Euclidean distance, pre-processing techniques like scaling or standardization are frequently used.

This measure may not always be in line with chemical or analytical relevance because it implies that all factors contribute equally to the distance. The procedure for performing agglomerative hierarchical clustering (AHC) in Excel is summarized step by step in Fig. 2.3.

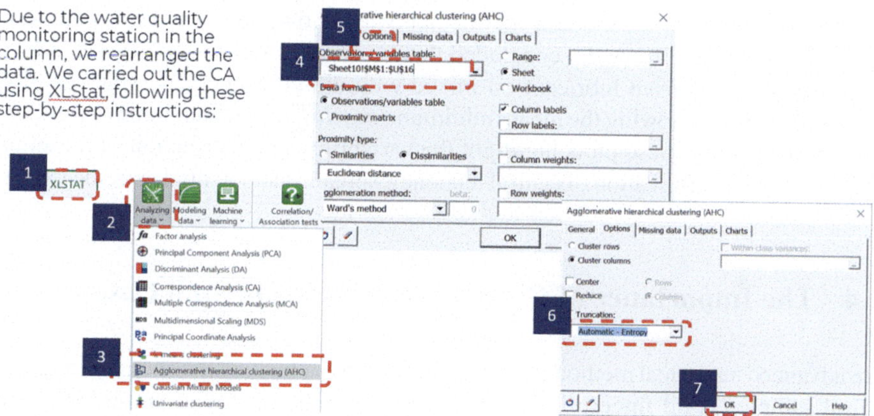

Fig. 2.3 Steps using Excel performing the AHC

2.3 Cluster Interpretation

- Cluster 1: The clustering of these several waste oils (MO, Bunker C, and diesel-based products) in this cluster suggests that the sampling stations likely comprised of similar petroleum hydrocarbon pollutants or pollution sources, leading to similar chemical properties.
- Cluster 2: This cluster indicates that lubricating oil mixed with diesel-based experience dominance in Cluster 2. It is possibly due to similar sources of petroleum hydrocarbon based, intentionally or unintentionally discharged at the sampling stations.
- Cluster 3: This cluster shows similar factors likely influence diesel-type product, affecting their chemical properties. Cluster 2 and Cluster 3 profile plots indicated the nearly look-alike fingerprint with different characteristics in the hydrocarbon-fuel alkanes fragmentation shares particular commonalities, while the difference of diesel-based product from Cluster 3 and Cluster 2 is due to significant differences in petroleum hydrocarbon oil concentrations of the n-alkanes compounds.

The grouping of oil spill samples was plotted in a variety of graphs other than dendrogram such as box-and-whisker plots for mean, minimum, maximum, and maximum-minimum ratio of diesel oil for Cluster 3 Box plots (Fig. 2.4), depending on the kind of data and the insights. In order to further examine the significant differences of the diesel-based product for the mean, minimum, maximum, and maximum-minimum ratio of diesel oils for Cluster 3 (GP6_Jetty) and Cluster 2 (PP3_liquid), respectively.

For the diesel type in Cluster 3, the highest concentration was 17,882.715 μg/ml, while the median was 2,463.238 μg/ml. The mean is 5,091.890, and the third quartile is 9,010.508. The profile was clearly at odds with the diesel in Cluster 2, where the mean was 49.886 and the median was 18.696 with the third quartile at 106.675. Therefore, it may be concluded that, in contrast to Cluster 2 diesel, which most likely comes from lubricant oil mixed with diesel. Figure 2.4 indicates box-and-whisker plots showing the mean, minimum, maximum, and maximum-minimum ratio of diesel oil. These plots highlight the variability and distribution of diesel oil data, allowing identification of central tendency, spread, and potential extreme values.

2.4 The Importance of Cluster Analysis in Oil Spill Case

An advanced analytical method for determining and contrasting the chemical makeup of oil samples is oil fingerprinting. By identifying the source of an oil leak, it is essential to environmental forensics and spill response. Analysis of the oil's distinct chemical "signature" is part of the procedure, with a primary focus on hydrocarbon profiles such as saturated hydrocarbons, or alkanes. Straight-chain hydrocarbons reveal information about the origin and stage of oil breakdown. Whereas PAHs

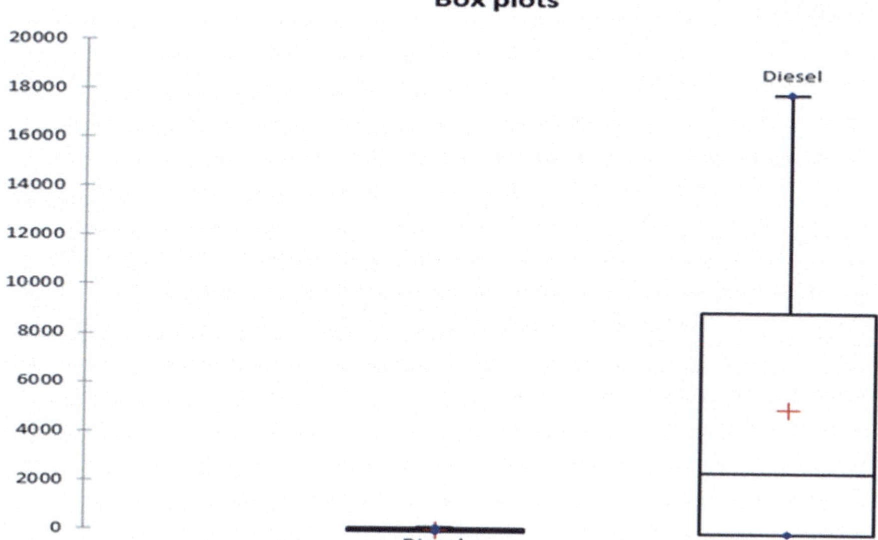

Fig. 2.4 Box-and-whisker plots for mean, minimum, maximum, and maximum-minimum ratio of diesel oil

and other aromatic hydrocarbons can act as indicators for the type of oil because they are frequently unique (Melo et al. 2022). The intricate chemical compounds in biomarkers that are stable throughout time provide accurate identification. They are derived from ancient biological material.

The oil fingerprinting benefits greatly from cluster analysis; the main reason is that by classifying oil samples with comparable chemical makeup, pattern recognition allows researchers to track the changes due to weathering or such. By identifying the source of an oil leak, it is essential to environmental forensics and spill response. Analysis of the oil's distinct chemical "signature" is part of the procedure, with a primary focus on hydrocarbon profiles. Differentiation of sources helps assign blame in situations with several spills or mixed oil kinds by differentiating between origins. Moreover, results of repeated sampling throughout time and clustering demonstrate how weathering, oxidation, and biodegradation alter the oil's chemical makeup. Importantly, legal proceedings, regulatory compliance, and environmental restoration initiatives are all aided by the insights obtained by oil fingerprinting employing clustering techniques.

References

Asif Z, Chen Z, An C, Dong J (2022) Environmental impacts and challenges associated with oil spills on shorelines. J Mar Sci Eng 10:762. https://doi.org/10.3390/jmse10060762

Kannel PR, Lee S, Kanel SR, Khan SP (2007) Chemometric application in classification and assessment of monitoring locations of an urban river system. Anal Chim Acta 582:390–399. https://doi.org/10.1016/j.aca.2006.09.006

Melo PTS, Torres JPM, Ramos LRV, Fogaça FHS, Massone CG, Carreira RS (2022) PAHs impacts on aquatic organisms: contamination and risk assessment of seafood following an oil spill accident. An Acad Bras Cienc 94:1215. https://doi.org/10.1590/0001-3765202220211215

Omran MGH, Engelbrecht AP, Salman A (2007) An overview of clustering methods. Intell Data Anal 11:583–605. https://doi.org/10.3233/ida-2007-11602

Templ M (2023) Visualization of missing values. In: Templ M (ed) Visualization and imputation of missing values: with applications in R. Springer, Cham, pp 107–150. https://doi.org/10.1007/978-3-031-30073-8_4

Chapter 3
Dimensionality Reduction of Oil Spill Variables

Abstract This chapter discusses the implementation of Discriminant Analysis (DA) as a statistical method for reducing the dimensionality and classifying oil spill data according to the chemical properties. DA is a group of oil samples by their chemical composition in order to identify the pollution sources. DA is one of the linear combination input variables, and its effective to be used in oil spill fingerprinting. The readers are able to practice step-by-step analysis of DA using Excel in this chapter, which consist stepwise forward and backward mode. After that, the readers will gain the new knowledge that DA is a powerful tool for dimensionality reduction and classification of oil samples, also enhancing the accuracy of source identification, especially in environmental forensic studies. A statistical method called Discriminant Analysis (DA) was used to group data according to certain characteristics. By examining the correlation between independent variables (features) and dependent variables (group labels), it develops a model that delineates the borders between various groups. Finding a set of discriminant functions or linear feature combinations that optimize the distance between groups is the primary objective. The model was used to forecast the group for recently acquired or unclassified data after trained. In disciplines including biology, business, and environmental science, DA is frequently employed for tasks like source of pollutants identification, market segmentation, and pollutant source tracing.

Keywords Discriminant analysis · Stepwise forward · Stepwise backward · Dimensionality · Pollutant · Source identification

3.1 Reduction of Dimensionality Using Discriminant Analysis (DA)

This technique can help in identifying the source of oil in maritime environments. This process can be completed by comparing the chemical composition of spilled oil samples, called as sources. The technique makes it easier to categorize and distinguish against different sources of oil pollution; the technique was very useful for oil spill

A. Ismail and H. Juahir, *Advanced Chemometrics*,
SpringerBriefs in Environmental Science, https://doi.org/10.1007/978-3-032-10742-8_3

fingerprinting. By examining indicators like the hydrocarbon profile (e.g., polycyclic aromatic hydrocarbons, or PAHs), it distinguishes between different types of oil, such as crude oil, diesel, and bunker oil, and their unique chemical signatures. Finding the canonical variables that optimize group separation was made easier with the use of Canonical Discriminant Analysis (CDA) (Abdullah et al. 2021).

This discriminant technique is a primary application, which divide into a set of observations or clustered data to numerous pre-established groups. Using this technique, the data can be clustered or grouped based on input parameters, called as predictors. A collection of discriminant functions (DFs), which are linear functions of the predictors, produced by this technique. The chemical "fingerprints" from possible sources, including different oil spillages and ship spills, were used to classify oil samples.

An effective analytical method for determining the chemical components causing geographic variance in oil spill data was the combination of GC-FID and GC–MS. The numerical evidence required to distinguish between the three distinct patterns found in the dataset was supplied by these strategies. This set of observations is sometimes referred to as the training set. Based on the training set, the technique constructs a set of linear functions of the predictors, known as discriminant functions, such that

$$L = b_1X_1 + b_2X_2 + \cdots + b_nX_n + C \tag{3.1}$$

where b is the discriminant coefficient, x is the input variables or predictors, and C is a constant. Further, the DA functionality aids in determining the best cluster, where n is the number of parameters used to classify a batch of data into a specific group, wij is the weight coefficient allocated by the discriminant factor (DF) to a given parameter (pij), i is the number of groups (G), and ki is the constant coherent to each group. Three modes—standard mode, forward stepwise mode, and backward stepwise mode—were used in DA to calculate the DF. The oil leak variables are added one at a time in the forward stepwise mode until no discernible changes are seen.

$$f(G_i) = k_i + \sum_{j=1}^{n} \text{wij.pij} \tag{3.2}$$

DA distinguishes between two or more groups (clusters) based on the characteristics related to the oil spill. With or without standardization in the construction of the discriminant factor (DF), the applied DA on the original dataset provides a comparable discriminant ability to the original dataset. DA separates two or more groups (clusters) according to the traits associated with the oil spill. The applied discriminant factor (DA) on the original dataset offers a discriminant ability that is comparable to the original dataset, regardless of whether the DF construction is standardized.

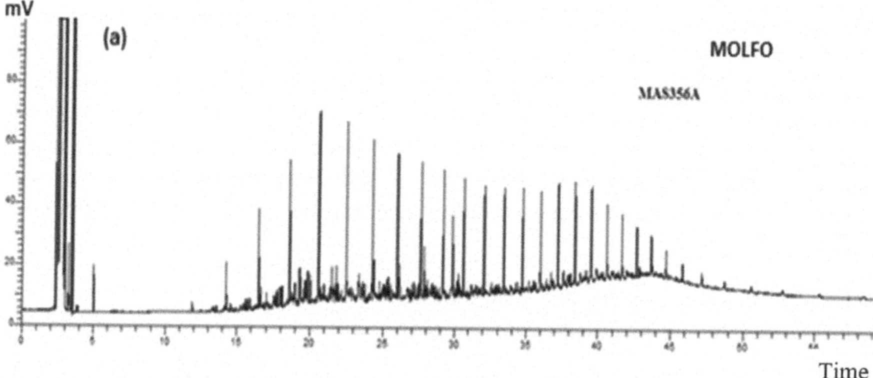

Fig. 3.1 Chromatographic profiling signatures distinguish by n-alkanes distribution profiles of MOLFO in oil samples

The one-leave-out approach in the backward stepwise mode was used to eliminate the oil spill variables, starting with the least significant and continuing until there are no noticeable changes. The oil spill data matrix (raw data) was used for DA; the independent variables are the oil spill parameters or compounds in the columns, and the dependent variables are in the rows. DA is a technique for classifying a set of observations into predefined classes. The purpose was to determine the class of an observation based on a set of parameters known as predictors or input variables. Gas chromatography-flame ionization detector (GC-FID) and gas chromatography-mass spectrometry (GC–MS) were chromatogram profiling for various types of oils. It distinguishes by its distribution of n-alkanes are effective analytical methods that are widely applied to the identification and measurement of chemical substances. The model was built based on a set of observations for which the classes are known. The chromatographic profiling signatures obtained by GC-FID and GC–MS show the n-alkane distribution profiles (C_{10}–C40) of MOLFO in oil samples as demonstrated in Fig. 3.1. The hydrocarbon distribution patterns indicate the significant difference in compositional variations across carbon number ranges.

3.2 DA as a Tool to Discriminate Patterns Using XLSTAT

This technique is also known as a supervised pattern recognition technique where the independent (x) and dependent (*y*) variables are taken into consideration. To determine the most important chemical compounds, numerical evidence is essential in addition to qualitative evaluations (Zali et al. 2021). This data demonstrates how their concentration values vary according to three different patterns. These trends might line up with particular situations, such as various oil spill incidents, particular environmental circumstances, or shifts in time. The chemical substances that are most responsive or suggestive of these patterns can be identified by examining these numerical changes.

Compounds that exhibit regular and statistically significant changes in concentration throughout the three patterns, for instance, were used as indicators for additional environmental evaluations or classification. By relating the found patterns to actual occurrences, these revelations not only improve the Discriminant Analysis model's interpretability but also reinforce its validity. The step-by-step to run the analysis can refer to Fig. 3.2.

1. Firstly, click the Add-ins in Excel, then XLSTAT.
2. Next, click the Discriminant Analysis (Fig. 3.2).

On the toolbar, click the Options, then for Qualitative select the range of data from the dataset. Likewise, the Quantitative by selecting the range of the dataset.

Oil spill variables were eliminated using the one-leave-out method in the backward stepwise manner, beginning with the least important ones and continuing until no discernible changes were seen. The most pertinent variables were kept in the dataset for analysis thanks to this procedure. For Discriminant Analysis (DA), the raw oil spill data matrix was used. The dependent variables in this matrix are arranged in rows, whereas the independent variables match the oil spill parameters or compounds shown in the columns. One statistical method used to group observations into predetermined classifications is Discriminant Analysis. Its main objective is to forecast an observation's class membership using a collection of factors known as predictors or input variables.

Oil spill variables were eliminated using the one-leave-out method in the backward stepwise manner, beginning with the least important ones and continuing until no

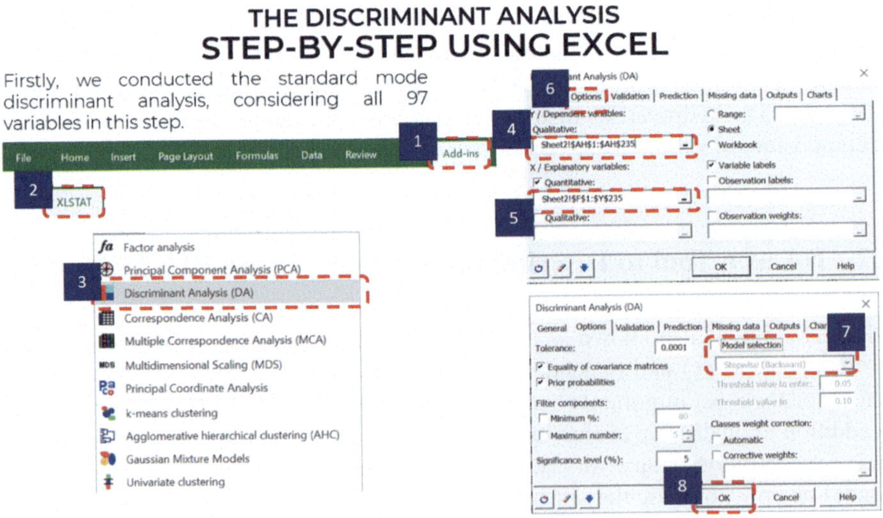

Fig. 3.2 Steps of discriminant analysis (DA) standard mode

discernible changes were seen. The most pertinent variables were kept in the dataset for analysis as part and parcel to this procedure. In this book, the raw oil spill data matrix was used. The dependent variables in this matrix are arranged in rows, whereas the independent variables match the oil spill parameters or compounds shown in the columns. In order to enable the algorithm to find patterns and correlations between variables, the model was built using a training dataset in which class membership was already known as illustrated clearly in Fig. 3.3. As shown in Fig. 3.3, Discriminant Analysis has identified significant parameters through consecutive modes of standard, stepwise forward, and stepwise backward.

As illustrated in Fig. 3.4, the stepwise DA and backward mode identified the most significant variables for group separation.

After following the steps, the arrangement of the results would be as in Table 3.1, which comprises standard mode, step forward mode, and step backward mode. An effective analytical method for determining the chemical components causing geographic variance in oil spill data was the combination of GC-FID and GC–MS. The numerical evidence required to distinguish between the three distinct patterns found in the dataset was supplied by these strategies.

With a Lambda value of 0.00 ($p < 0.0001$) from Wilk's Lambda test, the alternative hypothesis (Ha) is accepted (Ibrahim et al. 2023). Since the computed p-value is below the significance level of alpha 0.05, the alternative hypothesis ought to be accepted and the null hypothesis ought to be rejected (Fig. 3.5). Additionally, the forward stepwise mode DA separated the oil spill into four clusters/categories with a classification matrix accuracy of 100%. Both forward and backward DA stepwise analyses revealed that diesel, HFO, MOLFO, and WO were the yield of the analysis. Figure 3.5 displays the p-value of stepwise forward and backward, exhibiting the

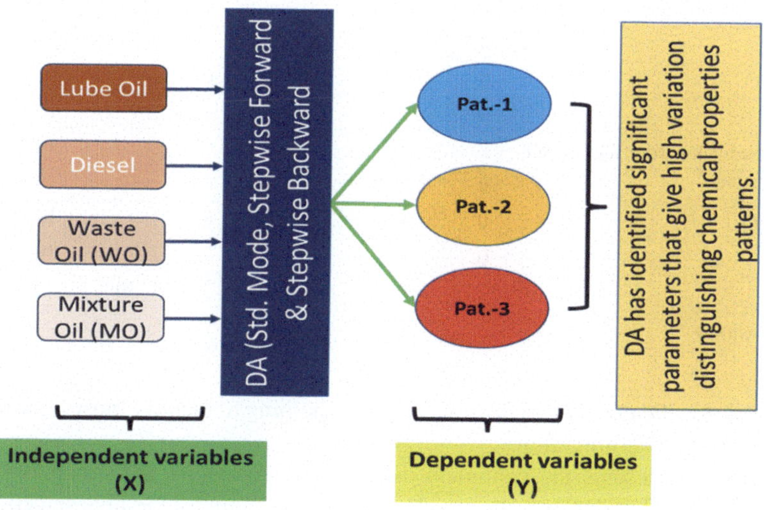

Fig. 3.3 Discriminant Analysis representing consecutive modes of standard, stepwise forward, and stepwise backward

MOD 'STEPWISE FORWARD/BACKWARD'

Secondly, we conducted the stepwise forward/backward discriminant analysis. We repeated all the previously shown steps with additional modifications in step 7, where we selected the "Model Selection" option.

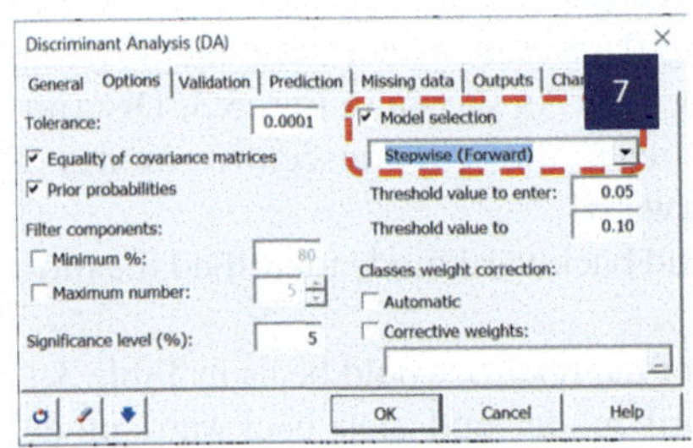

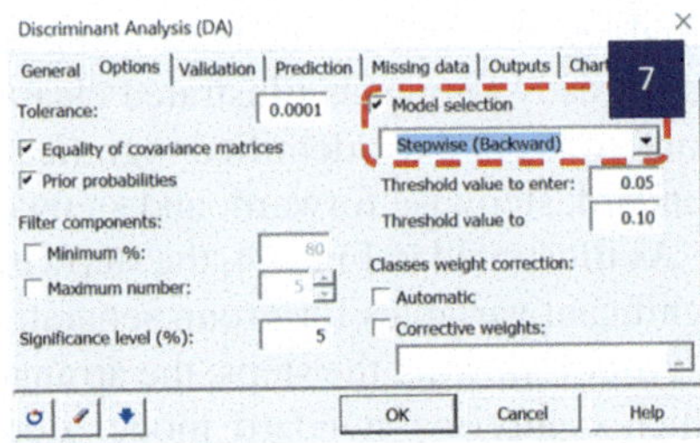

Fig. 3.4 DA with stepwise forward and backward mode

Table 3.1 Classification matrix of DA for all data measurements in oil classification

	Predicted types of oil by DA					
Types of oil	Diesel	HFO	MLFO	WO	Total	% Correct
Standard mode of DA diesel	2	0	0	0	2	100
HFO	0	12	0	0	12	100
MOLFO	0	0	9	0	9	100
WO	0	0	0	7	7	100
Total stepwise forward of DA	2	12	9	7	30	100
Diesel	2	0	0	0	2	100
HFO	0	12	0	0	12	100
MOLFO	0	0	9	0	9	100
WO	0	0	0	7	7	100
Stepwise backward of DA diesel	2	0	0	0	2	100
HFO	0	12	0	0	12	100
MOLFO	0	0	9	0	9	100
WO	0	0	0	7	7	100
Total	2	12	9	7	30	100

Source Author

statistical significance of variables at each step, helping to identify those contributing most to group discrimination.

Figure 3.6 presents the canonical discriminant functions and group centroids, highlighting the separation between classes without stepwise variable selection.

DISCRIMINANT ANALYSIS
RESULTS INTERPRETATION

Stepwise Forward DA

Lambda	F	DF1	DF2	p-value
0.439	147.486	2	231	< 0.0001
0.582	82.843	2	231	< 0.0001
		2	231	
		2	231	
0.908	11.665	2	231	< 0.0001
0.682	53.922	2	231	< 0.0001
		2	231	
		2	231	
0.961	4.709	2	231	0.010
0.884	15.087	2	231	< 0.0001
		2	231	
		2	231	
		2	231	
		2	231	
		2	231	
		2	231	
		2	231	
		2	231	
		2	231	
		2	231	

Stepwise Backward DA

Lambda	F	DF1	DF2	p-value
0.439	147.486	2	231	< 0.0001
0.582	82.843	2	231	< 0.0001
0.602	76.221	2	231	< 0.0001
		2	231	
0.908	11.665	2	231	< 0.0001
0.682	53.922	2	231	< 0.0001
		2	231	
0.960	4.855	2	231	0.009
		2	231	
0.884	15.087	2	231	< 0.0001
		2	231	
		2	231	
		2	231	
		2	231	
		2	231	
		2	231	
		2	231	
		2	231	
		2	231	
		2	231	

Fig. 3.5 p-value stepwise forward and backward Discriminant Analysis

DISCRIMINANT ANALYSIS
RESULTS INTERPRETATION

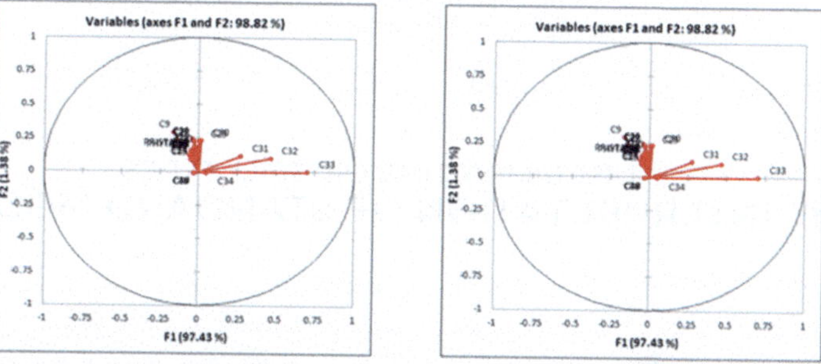

The plot illustrates the standard mode of discriminant functions for petroleum hydrocarbons parameters attributed to the type of oil spills due to seawater contamination categorized as Diesel, Heavy Fuel Oil (HFO), Mixed Oil Lubricant Fuel Oil (MOLFO) and Waste Oil (WO).

Confusion matrix for the estimation sample

Standard mode of DA						
DIESEL	2	0	0	0	2	100.00%
HFO	0	12	0	0	12	100.00%
MOLFO	0	0	9	0	9	100.00%
WO	0	0	0	7	7	100.00%
Total	2	12	9	7	30	100.00%

Fig. 3.6 Result of standard mode Discriminant Analysis (DA). *Source* Authors

Figure 3.7 shows the canonical discriminant functions and group centroids obtained using variables selected through stepwise selection, demonstrating enhanced discrimination performance relative to the standard DA mode.

In Fig. 3.8, the plot displays the canonical scores for each case, illustrating the grouping pattern and degree of overlap between the categories when no stepwise variable selection is applied.

DISCRIMINANT ANALYSIS

RESULTS INTERPRETATION

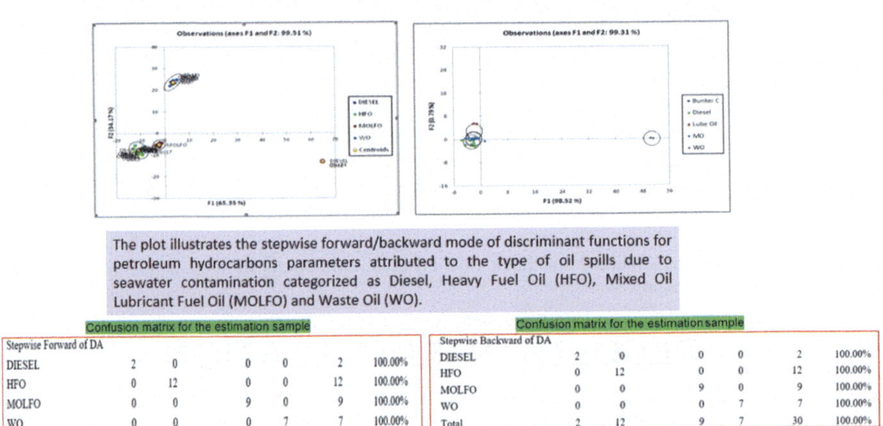

Stepwise Forward of DA						
DIESEL	2	0	0	0	2	100.00%
HFO	0	12	0	0	12	100.00%
MOLFO	0	0	9	0	9	100.00%
WO	0	0	0	7	7	100.00%

Stepwise Backward of DA						
DIESEL	2	0	0	0	2	100.00%
HFO	0	12	0	0	12	100.00%
MOLFO	0	0	9	0	9	100.00%
WO	0	0	0	7	7	100.00%
Total	2	12	9	7	30	100.00%

Fig. 3.7 Result of stepwise forward/backward mode discriminant analysis (DA)

DISCRIMINATING OIL SPILL TYPES

THE INTERPRETATION OF STANDARD MODE DA

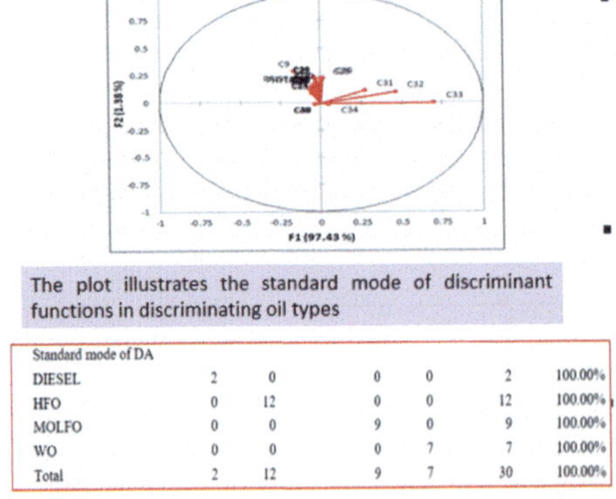

Standard mode of DA						
DIESEL	2	0	0	0	2	100.00%
HFO	0	12	0	0	12	100.00%
MOLFO	0	0	9	0	9	100.00%
WO	0	0	0	7	7	100.00%
Total	2	12	9	7	30	100.00%

- Results show that the DA gives 100% correct discrimination classification of predefined classes Cluster Analysis gives. On the other hands we can says that the 97 independent variables involve in this study for distinguishable types of oil spill within the study area are significantly plays important roles.
- However, it is essential to know the most significant until the least considerable variables are present to reduce the number of variables for further analysis within a short time execution and cost-effective in future sampling plans.

The Standard mode only considered the most significant variables as discriminant functions to discriminate the predefined classes.

Fig. 3.8 Plot of data interpretation of standard mode discriminant analysis (DA). *Source* Authors

DISCRIMINATING OIL SPILL TYPES
THE INTERPRETATION STEPWISE FORWARD OF DA

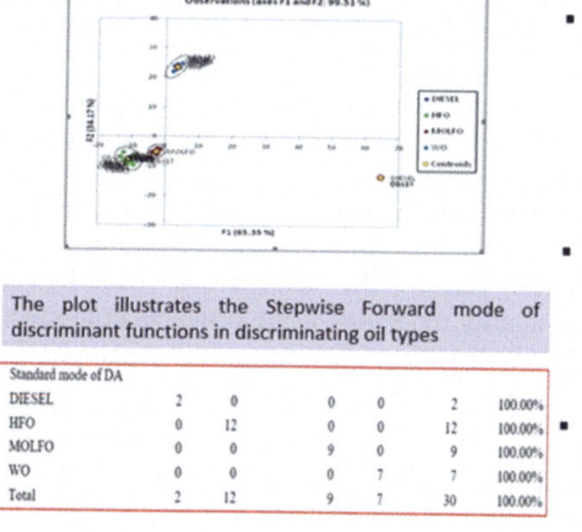

Standard mode of DA						
DIESEL	2	0	0	0	2	100.00%
HFO	0	12	0	0	12	100.00%
MOLFO	0	0	9	0	9	100.00%
WO	0	0	0	7	7	100.00%
Total	2	12	9	7	30	100.00%

- Results show that the stepwise forward DA gives 100% correct discrimination classification of predefined classes by Cluster Analysis. Similarly of the standard mode, from 97 independent variables involve in stepwise forward for distinguishable types of oil spill within the study area are significantly plays important roles.
- Stepwise forward DA was successfully discriminated the hydrocarbon oil spill. It demonstrated unquestionably that this method was effective in locating the discriminating factors in the oil's spatial variance.
- Again, only the most significant variables were considerable as discriminant functions to discriminate the predefined classes.

Fig. 3.9 Plot of stepwise forward mode discriminant analysis (DA) data interpretation. *Source* Authors

DISCRIMINATING OIL SPILL TYPES
THE INTERPRETATION OF STEPWISE BACKWARD DA

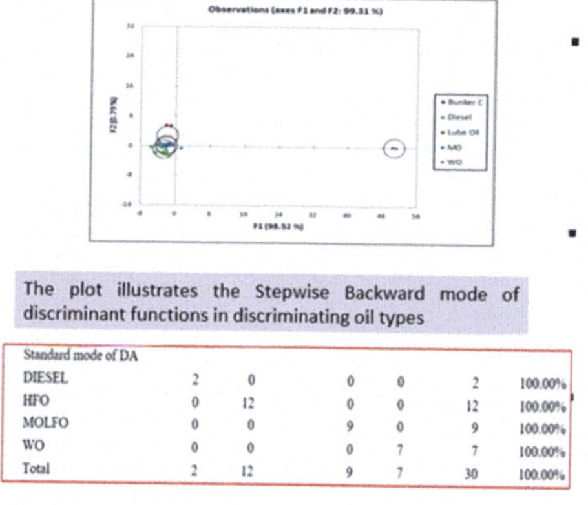

Standard mode of DA						
DIESEL	2	0	0	0	2	100.00%
HFO	0	12	0	0	12	100.00%
MOLFO	0	0	9	0	9	100.00%
WO	0	0	0	7	7	100.00%
Total	2	12	9	7	30	100.00%

- Results show that the step backward DA gives 100% correct discrimination classification of predefined classes Cluster Analysis gives. It was suggested that the 97 independent variables involve in step backward for distinguishable types of oil spill within the study area are significantly plays important roles.
- Similarly, stepwise backward DA was successfully discriminated the hydrocarbon oil spill. Again, it demonstrated unquestionably that this method was effective in locating the discriminating factors in the oil's spatial variance.

Likewise, only the most significant variables were considerable as discriminant functions to discriminate the predefined classes.

Fig. 3.10 Plot of data interpretation stepwise backward mode discriminant analysis (DA). *Source* Authors

Figure 3.9 illustrates the canonical scores and group centroids based on variables selected through stepwise forward selection, highlighting improved separation and reduced overlap between groups relative to the standard DA mode.

Figure 3.10 shows the canonical scores and group centroids obtained from variables selected through backward stepwise elimination, allowing comparison with the forward stepwise approach to evaluate the robustness of group separation.

References

Abdullah SNF, Abdullah N, Ishak R, Hasan N, Wahab WABA, Zakaria NA, Zainuddin N (2021) Estimation of rainwater harvesting by the reflectance of the purity index of rainfall. Environ Sci Pollut Res 28:35613–35627. https://doi.org/10.1007/s11356-021-12772-6

Abdul Zali M, Juahir H, Aris AZ, Toriman ME, Mokhtar M, Zainuddin SFA, Yunus K, Gazim MB, Kamarudin MKA, Kasan NA, Khalit SI, Aizat WM (2021) Tracing sewage contamination based on sterols and stanols markers within the mainland aquatic ecosystem: a case study of Linggi catchment, Malaysia. Environ Sci Pollut Res 28:20717–20736. https://doi.org/10.1007/s11356-020-11680-5

Ibrahim A, Juahir H, Toriman ME, Kamarudin MKA, Gasim MB, Saudi ASM, Umar R, Che Hasnam CN, Aziz NAA, Latif MT (2023) Water quality modelling using principal component analysis and artificial neural network. Mar Pollut Bull 187:114493. https://doi.org/10.1016/j.marpolbul.2022.114493

Chapter 4
Contaminant of Polyaromatic Hydrocarbon in Oil Spills

Abstract This chapter highlights the application of Principal Component Analysis (PCA) in order to analyze oil spill contaminants, specifically focusing on PAH as a source identification and dimensionality reduction. This chapter described the dominant linear combinations of original variables that explain variability, filter out less significant parameter while retaining essential data structure. Plus, this chapter also simplifies the oil spill sample datasets which derived from GC-FID and GC–MS methods in order to get better performance in data visualization, noise reduction and also machine learning performance. As conclusion in this chapter shown that PCA is a powerful and essential tool in oil spill fingerprinting. By emphasizing significant PAH patterns, it improves the clarity of chemical datasets and supports environmental forensics through effective source tracing. Principal Component Analysis (PCA) is known as one of the popular techniques for reducing dimensionality in diverse field, including data science and statistics. This technique enables breaking down complicated datasets into a collection of uncorrelated variables known as main components, which represent the greatest amount of variance in the data. Moreover, it has since grown to be a vital tool in exploratory data analysis (Karl in Structure and properties of condensed matter 2:11 (1901), especially for finding correlations, patterns, and trends in high-dimensional datasets. A linear combination of the initial variables makes up each main component in PCA. The largest variance in the data is captured by the first Principal Component, which is followed by the second component, which represents the next biggest variance, and so forth. By applying this technique, the complicated data derived from oil mixtures is made simpler, making the dataset easier to handle and understand. By this procedure, the technique aids in identifying and removing the most important factors that contribute to the variations among the oil samples, making it easier to understand the data in a way that is more relevant and clearer. Assume we have thousands of photos in a massive photo album. Color, brightness, objects, and other details are all present in every image. What would happen, though, if minimization the file size while preserving the most crucial information? Principal Component Analysis, or PCA, is what does that. It allows to preserve the important details while compressing, eliminating noise, visualizing complex data, and enhancing machine learning models. Specific objectives of PCA: • An objective of this technique is to identify linear combinations of the original variables that are

A. Ismail and H. Juahir, *Advanced Chemometrics*,
SpringerBriefs in Environmental Science, https://doi.org/10.1007/978-3-032-10742-8_4

useful in accounting for the variation in those variables. This is effectively a clustering of the variables. • It provides information on the most meaningful parameters due to spatial and temporal variations that describe the whole dataset by excluding the less significant parameters with minimum loss of original information.

Keywords Principal component analysis · High-dimensional · Variance · Oil mixtures · Uncorrelated variables

4.1 Principal Component Analysis in Oil Spill Fingerprinting

PCA preserves the most important information for reducing the dimensionality of the oil spills dataset. By using this technique, the complicated data derived from GC-FID and GC–MS is made easier to handle and understand. A better and more relevant interpretation of the data is made possible by the technique's assistance in highlighting and extracting the most important variables that contribute to the variations within the oil samples. The amount and presence of various polycyclic aromatic hydrocarbons (PAHs) in oil samples are evaluated using PCA in order to determine the source of contamination. This facilitates the tracing and attribution of oil spills to particular sources, which is crucial for remediation and environmental forensics (Kannel et al. 2007). This technique is essential for determining the chemical signatures of various oil samples. This shows how important it can be used in actual environmental situations and connects it to oil spill fingerprinting. Oil spills generate intricate environmental data with many interconnected variables, which is consistent with how PCA makes high-dimensional datasets simpler. Its value in source identification is a crucial component of environmental forensics as highlighted by its ability to disclose underlying patterns and facilitate the distinction between different sources of oil contamination. The roles of PCA in oil spill fingerprinting from the laboratory analysis of gas chromatography-flame ionization detector (GC-FID) and gas chromatography-mass spectrometry (GC–MS) results are succinctly illustrated in Fig. 4.1.

4.2 Correlation Between Each Original Variable and a Principal Component

This technique was to form new variables which are linear deposits of the original variables. The maximum number of variables that can be formed is equal to the number of the original variables, and the new variables are uncorrelated among them. PCA techniques were used to identify the most significant variables that determine the oil type classification variables by excluding the least significant with the most

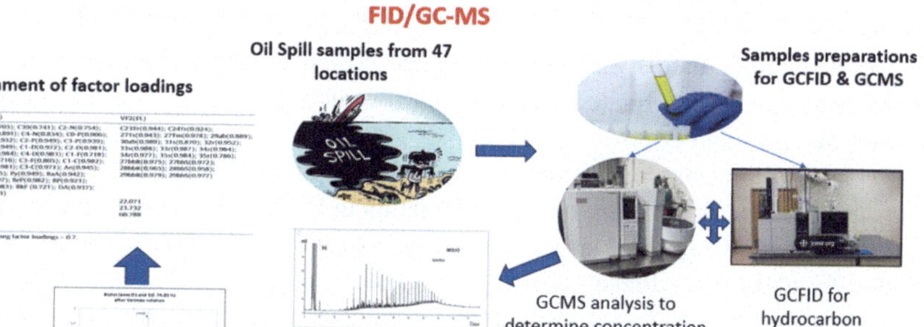

Fig. 4.1 Roles of PCA in oil spill fingerprinting

significant without losing the original information (Kannel et al. 2007). The most crucial information (variance) from the original dataset is captured by a Principal Component (PC), a new feature (or dimension) produced by PCA. PCA constructed the latent variables (scores) which were the linear combinations of the original data of oil spill compounds. In the analysis, only PCs with eigenvalues greater than 1.0 were only considered significant. The following equation (Kannel et al. 2007) was used in the analysis of Principal Component (PC):

$$Z_{\mathrm{ij}} = a_{i1}X_{1j} + a_{i2}X_{2j} + \ldots + a_{\mathrm{im}}X_{\mathrm{mi}} \tag{4.1}$$

where z denoted as the component score, a as the component loading, x as a variable with measured value, i as component number, j as sample number, and m as the total number of variables.

The application of this technique can reduce the complex mixture of oil data and obtain a useful data interpretation of the most significant discriminating variables. Moreover, it allows the complex mixture data extracted from GC-FID and GC–MS to be classified which is not usually achievable through conventional fingerprints. Reducing the dimensionality of the complex GC–MS dataset while preserving the most significant variance in the petroleum hydrocarbon composition allows for source identification and weathering trend analysis.

4.3 PCA as a Tool to Pattern Recognition Using XLSTAT

PCA is probably the most widespread multivariate statistical technique used in chemometrics, and because of the importance of multivariate measurements in chemistry, it is regarded by many as the technique that most significantly changed the chemist's view of data analysis (Arslan et al. 2023). These elements help to preserve the most significant information while eliminating superfluous or irrelevant aspects by capturing the greatest variance in the data. Principal Component Analysis (PCA), which simplifies complicated datasets and reveals significant patterns, is crucial to comprehending and controlling oil spills.

Principal Components are the new variables extracted from the original dataset that serves as the underlying main components in PCA throughout the analysis. The eigenvectors and eigenvalues of the data's covariance or correlation matrix serve as the foundation for these elements (Ibrahim et al. 2023). Principal Components provide a number of crucial roles in PCA analysis, helping to efficiently transform and evaluate the data to fit the following functionalities.

- Reduction of Data Dimensionality

Multiple variables, including chemical composition, environmental characteristics, and spatial distribution, are frequently included in oil spill datasets. PCA preserves the most important information while reducing the number of variables. This simplicity facilitates effective analysis and visualization, which makes it simpler to pinpoint the main variables affecting the behavior and effects of oil spills.

- Recognizing Variability in Chemical Composition

By breaking down oil samples' chemical complexity into its Principal Components, PCA reveals differences in important molecules such as aromatics, alkanes, and polycyclic aromatic hydrocarbons (PAHs). This makes it possible to comprehend the weathering and degrading processes of oil in greater depth.

- Citation of the Source

PCA can distinguish between different types of oil by examining its chemical fingerprints, such as hydrocarbon profiles (Corilo et al. 2013). This capacity is crucial for identifying the source of oil spills whether they come from a pipeline, a ship, or natural seepage and allocating blame. This approach allows in comprehending and controlling oil spills by the simplification of intricate datasets and the discovery of significant patterns; these revelations not only improve the Discriminant Analysis model's interpretability but also reinforce its validity.

4.4 Steps on How to Run the PCA

The step by step to run the analysis can refer to Fig. 4.2.

1. Firstly, click the Add-ins in Excel, then XLSTAT.
2. Next, click the Principal Component Analysis (Fig. 4.2).

On the toolbar, click the Options, then select the observations or variables table range of the oil spill dataset. For PCA type select Pearson, click Sheet and variable labels, then click OK.

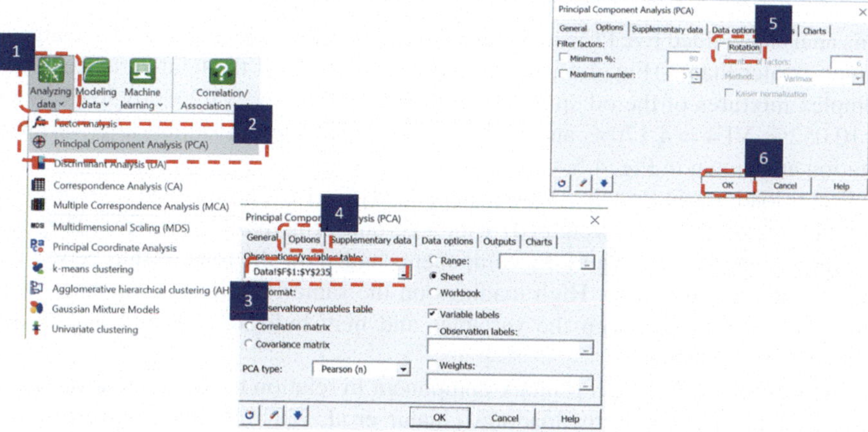

Fig. 4.2 Step by step using Excel to run PCA

Figure 4.3 illustrates the Principal Component Analysis (PCA) score plot for source identification with factor loadings.

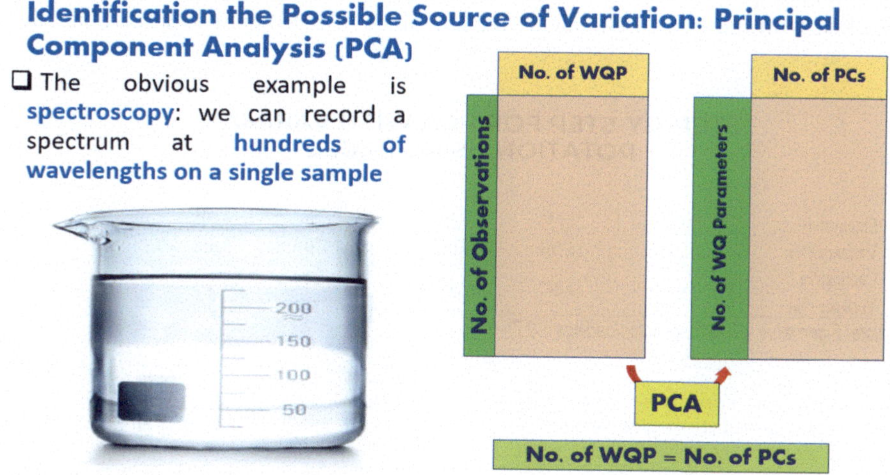

Fig. 4.3 PCA for various sources identification

4.5 Results of Principal Component Analysis

This analysis yielded five PCs, or Variance Factors (VFs), with mean data and eigenvalues greater than 1 (Fig. 4.4), accounting for 93.088% of the total variance in the complex mixtures of the oil spill dataset (VF1 = 58.449%, VF2 = 16.564%, VF3 = 10.052%, VF4 = 4.170%, and VF5 = 3.853%. The factor loadings after varimax rotation are shown in Fig. 4.5.

Factor loadings in Principal Component Analysis (PCA) show how each original variable is correlated (or weighted) with a certain Principal Component. They aid in interpreting the meaning of the components and show the relationship between a variable and a component. High loadings on the same Principal Component indicate a relationship between the variables and may indicate a shared underlying characteristic or factor.

The significance of each primary component in relation to the original variables can be deduced using factor loadings (Juahir et al. 2017). Certain n-alkanes and polycyclic aromatic hydrocarbons (PAHs) are examples of compounds with factor loadings more than 0.7 that have been found to be important indicators for identifying oil sources and describing weathering processes. While changes in lighter hydrocarbons reflected weathering trends throughout subsequent components, the high factor loadings (>0.7) of particular biomarkers and PAHs along PC1 demonstrated their dominance in explaining the variability associated with source-specific fingerprints. By eliminating noise from less important components and bringing the PCA results into line with the known chemical characteristics of petroleum hydrocarbons, the

PCA

STEP-BY-STEP FOR PCA WITH VARIMAX ROTATION USING EXCEL

Eigenvalue	50.850	24.255	7.596	6.640	4.005
Variance (%)	58.449	16.564	10.052	4.170	3.853
Cumulative Variance (%)	58.449	74.847	82.443	89.084	93.088

Note: Eigenvalue > 5; Strong factor loadings > 0.7

Secondly, we conducted the PCA with varimax rotation. We repeated all the previously shown steps with additional modifications in step 5, where we selected the "Rotation" option and fill in "Number of factors such as 5 or 7 (only rotate PCs with eigenvalue ≥ 1) and in the "Method" select "Varimax".

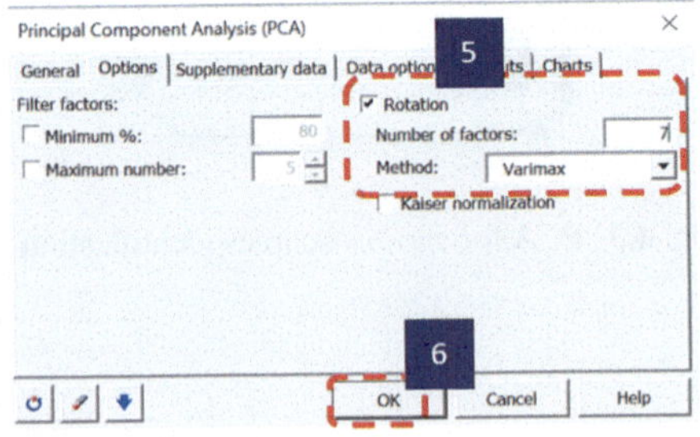

Fig. 4.4 Result of PCA with varimax rotation

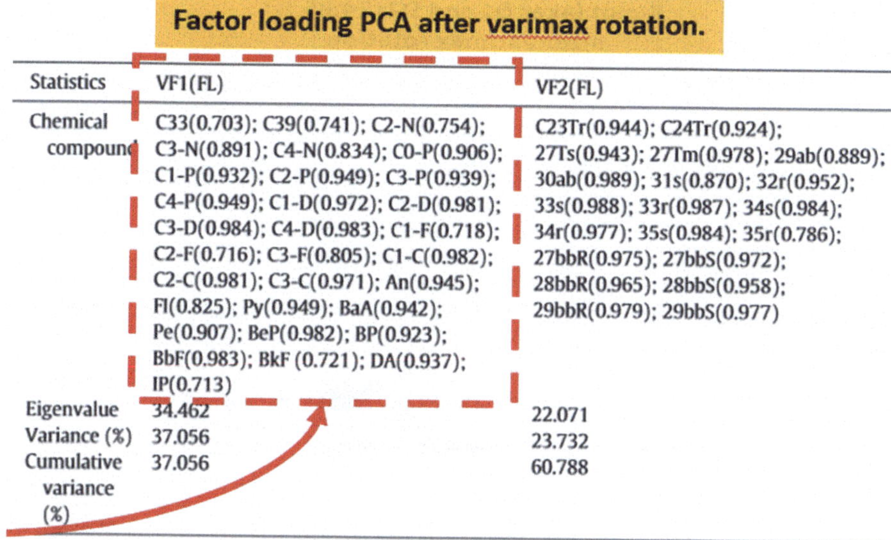

Factor loading PCA after varimax rotation.

Statistics	VF1(FL)	VF2(FL)
Chemical compound	C33(0.703); C39(0.741); C2-N(0.754); C3-N(0.891); C4-N(0.834); C0-P(0.906); C1-P(0.932); C2-P(0.949); C3-P(0.939); C4-P(0.949); C1-D(0.972); C2-D(0.981); C3-D(0.984); C4-D(0.983); C1-F(0.718); C2-F(0.716); C3-F(0.805); C1-C(0.982); C2-C(0.981); C3-C(0.971); An(0.945); Fl(0.825); Py(0.949); BaA(0.942); Pe(0.907); BeP(0.982); BP(0.923); BbF(0.983); BkF (0.721); DA(0.937); IP(0.713)	C23Tr(0.944); C24Tr(0.924); 27Ts(0.943); 27Tm(0.978); 29ab(0.889); 30ab(0.989); 31s(0.870); 32r(0.952); 33s(0.988); 33r(0.987); 34s(0.984); 34r(0.977); 35s(0.984); 35r(0.786); 27bbR(0.975); 27bbS(0.972); 28bbR(0.965); 28bbS(0.958); 29bbR(0.979); 29bbS(0.977)
Eigenvalue	34.462	22.071
Variance (%)	37.056	23.732
Cumulative variance (%)	37.056	60.788

Note: Eigenvalue > 5; strong factor loadings > 0.7.

Fig. 4.5 Factor loadings of PCA after varimax rotation

inclusion of variables with factor loadings greater than 0.7 guaranteed robust interpretation. Upon run the analysis, the Score Plot of PCA yielded the diesel, lube oil, MO, Bunker C, and WO representing the types of oil according to its intrinsic chemical properties as illustrated in Fig. 4.6.

4.6 Data Interpretation of Principal Component Analysis (PCA)

In this book, PCA was used to obtain a usable data interpretation of the most important discriminating characteristics and to decrease the complicated mixes of oil sample data. Complex mixtures of data extracted from GC-FID and GC–MS can now be identified, which was previously impossible with traditional fingerprints. Using mean data, PCA varimax factors (VFs) produced five variance factors that accounted for 93.09% of the overall variation in complicated oil spill mixes (Table 4.1). 50.592% of all oil spill components were caused by varimax factor (VF1) (Fig. 4.5). In Peninsular Malaysia, VF2 was accountable for 24.255% of the overall variation in oil spills. In the varimax rotation, eigenvalues larger than five were the primary criterion taken into account.

Following varimax rotation, the varimax factor (VF1) with the highest loading factor (50.85), as shown in Table 4.1, was found to have a strong correlation with the following variables of saturated hydrocarbon oils: n-alkanes (C10—C30),

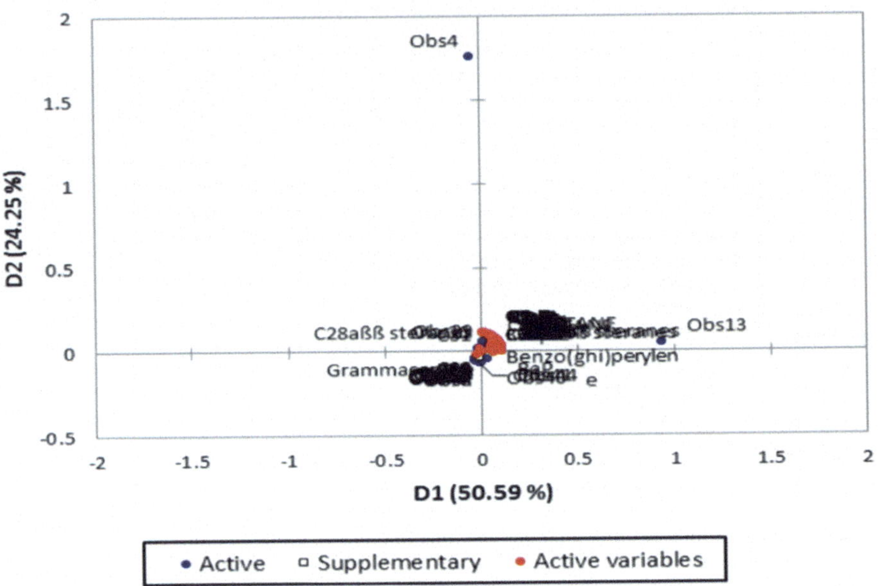

Fig. 4.6 Score plot of PCA, the oil samples in the analysis replicates of the diesel, lube oil, MO, Bunker C and WO. *Sources* Authors

isoprenoids (Phytane), n-alkanes (C0N—C4N), chyrene (C0C—C3C), other EPA priority PAH pollutants (Acl, Ace, BpH, An, FI, Py, BaA, Bph, Pe, BeP, BP, BbF, BkF, BaP, DA, and IP), and biomarkers (terpanes).

On the other hand, VF1 has a negative correlation with other EP priority PAH molecules, including C31R-C29aßß steranes. The deliberate release of products based on diesel and marine engines, such as lubricating oil, into the environment was the primary source of n-alkanes (C10–C30), isoprenoids Phytane, Naphthalene (C1–N), Chrysene (C0C), and other EPA priority PAH pollutants the seawater of Peninsular Malaysia. VF2 had a positive correlation with other alkylated PAH contaminants that were caused by the inadvertent or unlawful release of waste oil and petroleum products from shipping operations and fishing vessels. The hydrocarbon molecules in the oil complex mixture were dominated by PAHs with a greater molecular weight, ranging from four to six rings. It appears that VF3, VF4, and VF5 showed a strong link with a hydrocarbon oil compound of biomarker, indicating that waste oil from diesel and lubricant oil contributed significantly. According to biomarkers such as terpenes and steranes are often found as molecular markers in the environment, and their unique fragmentation complements environmental forensics' interest in identifying the source of oil spills.

A key component of forensic oil spill investigation, PCA improves the accuracy of source identification and environmental effect assessment while providing clarity

Table 4.1 Factor loadings PCA after varimax rotation for oil spill chemical compound

Statistics	VF1(FL)	VF2(FL)	VF3(FL)	VF4(FL)	VF5(FL)
Chemical compound	C10(0.990); C11(0.992); C12(0.995); C13(0.994); C14(0.970); C15(0.923); C16(0.904); C17(0.899); C19(0.884); Phytane(0.804); C19(0.868); C20(0.856); C21(0.840); C22(0.833); C23(0.823); C24(0.815); C25(0.832); C26(0.879); C27(0.906); C28 (0.922); C29 (0.917); C30 (0.876); C1-N(0.959); C0-C(0.800); C3-C(0.988); Bp(0.992); Acl(0.930); Ace(0.920); An(0.893); BaA(0.994); BbF (0.992); BkF (0.994); BeP (0.866); Pe (0.946); IP (0.982); DA (0.849); C23 (0.913); C24 (0.850); 27Ts (0.895); 27Tm (0.888); 29ab (0.875); 30ab (0.858);	C1-D (0.790); C2-D (0.974); C3-D (0.963); C4-D (0.860); C0-F (0.905); C1-F (0.989); C2-F (0.991); C3-F(0.989); C2-C (0.697); Fl (0.937); Py (0.959);	31S (0.733); Gammacerane (0.887); C28aßß steranes (0.798)	33R (0.703); 4S (0.951); 34R (0.950); 35S (0.944); 35R (0.951)	C31 (0.793); C32 (0.970)
Eigenvalue	50.850	24.255	7.596	6.640	4.005
Variance (%)	58.449	16.564	10.052	4.170	3.853
Cumulative Variance (%)	58.449	74.847	82.443	89.084	93.088

in complex information. The association between the Principal Components and the original variables (hydrocarbon compounds) is represented by factor loadings, which emphasize how each chemical contributes to the observed variation. Only variables with factor loadings higher than 0.7 were deemed significant in this investigation, guaranteeing that the interpretation concentrated on substances that had a considerable impact on sample separation along the major components.

References

Arslan N, Majidi Nezhad M, Heydari A, Astiaso Garcia D, Sylaios G (2023) A principal component analysis methodology of oil spill detection and monitoring using satellite remote sensing sensors. Remote Sens 15:1460. https://doi.org/10.3390/rs15051460

Corilo YE, Podgorski DC, McKenna AM, Lemkau KL, Reddy CM, Nelson RK, Rodgers RP, Marshall AG (2013) Oil spill source identification by principal component analysis of electrospray ionization Fourier transform ion cyclotron resonance mass spectra. Anal Chem 85:9064–9069. https://doi.org/10.1021/ac401604u

Ibrahim A, Juahir H, Toriman ME, Kamarudin MKA, Gasim MB, Saudi ASM, Umar R, Che Hasnam CN, Aziz NAA, Latif MT (2023) Water quality modelling using principal component analysis and artificial neural network. Mar Pollut Bull 187:114493. https://doi.org/10.1016/j.marpolbul.2022.114493

Juahir H, Ismail A, Mohamed SB, Toriman ME, Kassim AM, Zain SM, Ahmad WKW, Wah WK, Zali MA, Retnam A, Taib MZM, Mokhtar M (2017) Improving oil classification quality from oil spill fingerprint beyond six sigma approach. Mar Pollut Bull 120(1–2):322–332

Kannel PR, Lee S, Kanel SR, Khan SP (2007) Chemometric application in classification and assessment of monitoring locations of an urban river system. Anal Chim Acta 582:390–399. https://doi.org/10.1016/j.aca.2006.09.006

Karl P (1901) Structure and properties of condensed matter. Lond Edinb Dublin Philos Mag J Sci 2:11

Chapter 5
Solving Pattern Recognition Challenges Using Artificial Neural Networks (ANNs)

Abstract This chapter introduces the Artificial Neural Networks (ANNs) in chemometrics and also pattern recognition which is highlighted as a vital tool in chemometrics, particularly in analyzing the complex chemical datasets involved into nonlinear relationships. As described in this chapter, the authors choose to use ANNs because ANNs are applicable to manage high-dimensional, noisy and can use nonlinear data more efficiently than traditional methods. The implementation also explains by using gas chromatography-flame ionization detector (GC-FID) and gas chromatography-mass spectrometry (GC–MS) for classification. Artificial Neural Networks (ANNs) are a very helpful tool in chemometrics for analyzing complex chemical data, addressing problems with regression, classification, and pattern recognition (PR). Their ability to handle large datasets of oil spills and nonlinear interactions makes them particularly well-suited for chemometric applications. As is common in spectroscopy, chromatography, and other analytical techniques, ANNs play a key role in pattern recognition and data modeling in chemometrics. They are adapting at identifying underlying patterns in multi-dimensional and noisy chemical data. ANNs have become an important tool in the field of oil spill fingerprinting because of their ability to handle complex, nonlinear data and extract relevant patterns. This ability is essential in environmental forensics, where identifying the origin of oil spills is critical to reducing environmental damage and identifying the culprit. The Artificial Neural Networks (ANNs) are machine learning techniques and cognitive science that solve time-series forecasting issues, recognize patterns, and produce state-of-the-art outcomes for a wide range of tasks that are difficult to complete with rule-based programming, such as the voice recognition and computer vision. The ANNs are commonly defined as a biologically adaptive technique using the three sorts of parameters listed below: 1. The pattern of connections among the distinct layers of neurons. 2. The interconnectivity weight's learning procedure for updating. 3. The activation function, which is the method by which the weighted input of a neuron is transformed into the activation of its output. Complex hydrocarbon mixtures that differ based on the source (crude oil, refined products, etc.) are found in oil spills. To differentiate between various oil sources, ANNs can evaluate chemical composition data, such as gas chromatography-mass spectrometry (GC–MS) results. The ANN is trained using features including weathering effects, hydrocarbon ratios, and

A. Ismail and H. Juahir, *Advanced Chemometrics*,
SpringerBriefs in Environmental Science, https://doi.org/10.1007/978-3-032-10742-8_5

biomarker profiles to categorize samples according to their source. Anyhow, in this book oil spill datasets from GC-FID and GC–MS are used to run the ANNs.

Keywords Artificial neural networks · Pattern recognition · Cognitive science · Onlinear · Activation function

5.1 ANNs Track Procedures and Identify Irregularities in the Results of Laboratory Analysis

1. Integration and Processing of Data

To increase fingerprinting accuracy, ANNs can combine data from several analytical methods, such as spectroscopy and chromatography. They work effectively with real-world environmental samples because they are resilient to noisy or missing data (Li et al. 2021).

2. Process Observation

Data Consistency: By examining past data, ANNs can be trained to identify trends in laboratory operations. Any departure from these trends may indicate possible problems.

Parameter tracking: ANNs assist in guaranteeing compliance with established protocols by keeping an eye on crucial factors like temperature, pressure, and time throughout operations.

In reality, the tools were initially presented as an introductory training statistical guidance of the back-propagation method for ANN feed-forward, making it a significant inference system of nonlinear approaches to address the inaccuracy of multi-linear regression (MLR) techniques. Notably, it was specifically created for practical application in data mining and forecasting. Nowadays, the networks enable a broad range of applications with excellent forecast or prediction accuracy, speed, and precision. Computer elements relied on a large number of unknown inputs make up the network, numbered in the tens or hundreds. Additionally, the neural configuration networks were built using back-propagation training and a three-layered feed-forward, consisting of (a) input layers, (b) hidden layers, and (c) output layers (Fig. 5.1).

5.2 Steps to Run ANNs Application

To proceed with the ANN application, the following steps were adhered to:

1. Division of the Dataset

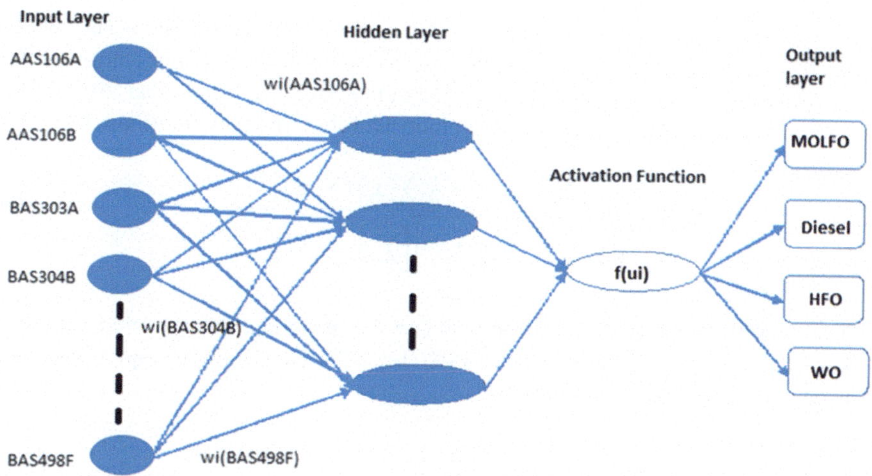

Fig. 5.1 Artificial Neural Network–multilayer perceptron (ANN-MLP) architecture model for oil type classification. *Source* Authors

A sizable oil spill dataset was separated into two parts before the ANNs could begin processing it: (a) the sets that would be used for additional ANN training, testing, and validation; and (b) the dataset that was available, which offered a valid-based selection parameter of the best input. To guarantee the prediction accuracy of ANNs, the dataset partition process was meticulously completed. The oil spilled dataset was validated using the validation set procedure in order to identify the so-called successful ANNs training. In order to establish the output value, the validation process used either known or unknown fresh input data from the oil spilled substances fed into the ANN. The data split in ANN was done at random using the input. According to the testing of the observations or dataset was only 10% of the entire dataset, 75% of the dataset for training process, and only 15% for validation. The method of selection input considers the various factors like ecological, linear correlation, methods of data mining, forward selection and backward elimination, and sensitivity analysis using trained ANN.

2. Learning Process

The learning process in ANN was commonly as a nonlinear optimization significant for the determination of any error function. Such ANN learning process was reflecting the key information ability process through the adaptive link weights (fully interconnected link weights) in identifying the best diagrammatic optimization of oil spilled input dataset. It enables the process to acquire the most correct output through learning. Besides, the learning process serves the error function minimization through the iterative technique. The iterative process was only terminated when the well-predefined value of oil spilled error function established (Ismail et al. 2023).

3. Determination of Model Performance (Validation)

a. The model performance determination or model validation functions were to determine the predictive adequacy of the model using the application of the residual analysis, to analyze and predict the errors (Schmidhuber 2015). The method to predict errors was to include the following computations (Eq. 5.1):

$$R = \frac{\left[\sum_{i=1}^{n} x_i x_i - \frac{1}{n}\left(\sum x_i\right)\left(\sum x_i\right)\right]^2}{\sqrt{\left[\sum x_1^2 - \frac{1}{n}\left(\sum x_i\right)^2\right]\left[\sum x_1 - \frac{1}{n}\left(\sum x_i\right)^2\right]}}. \tag{5.1}$$

b. The overall deviation or variation between the mean and the expected values is known as the Sum of Square Error (SSE) (Eq. 5.2). The following equations were used to describe the computation:

$$\text{SSE} = \sum_{i=1}^{n} (x_i + x_i)^2. \tag{5.2}$$

c. The following formula was used to express the Root Mean Square Error (RMSE) (Eq. 5.3):

$$\text{RMSE} = \sqrt{\frac{1}{n} \sum_{i=1}^{n} (x_i - x_i)^2}. \tag{5.3}$$

By presuming that the higher forecast error was significantly more significant than the smaller forecast error, RMSE precisely assesses the imperfection matches to the predictor and observed parameter or variable. It fits the oil spill scenario precisely if the RMSE is zero.

d. Percent Residual Error (%RE) is explicitly expressed as Eq. 5.4:

$$\%\ \text{RE} = \frac{x_i - x_i}{x_i} \times 100. \tag{5.4}$$

ANN methods applied in this study was restricted to the multilayer perceptron (MLP) as a substantial predictive and classification model of oil spill fingerprinting. The application of Artificial Neural Network (ANN-MLP) methods in this study is explicitly discussed in detail as the following;

(a) Multilayer perceptron (MLP) of Artificial Neural Network

Multilayer perceptron (MLP) of Artificial Neural Network as a high-dimensional clustering and prediction in data mining has become the famous technique application in classification and data analyses. The ANN-MLP was performed in this study using feed-forward propagation to classification-type and regression-type problems. The ANN-MLP model used in the oil spill fingerprinting serves a good range of classifier pattern for nonlinearity separable, and the inherent learning algorithm was reliable and provided good generalization whenever it dealt with the uncertainties of data and random events. In this study, the common ANN-MLP consists of an input layer, hidden layers, and one output layer. Hidden layers can be more than one. The input layer nodes or neurons associated with the oil spill variable number describe how the variable is classified. The number of output layers determined the number of clusters or classes (type of oil from the spilled sources) that equaled to the number of nodes. An ANN restricted MLP model was significantly employed in this study to predict the most precise classification and regression of oil spilled type according to the intrinsic physico-chemical properties.

The fact about an ANN is that, the number of nodes in the hidden layers and the output layers itself have a strong reliance with the complexity task of the classification and regression of oil spill compounds, where the substantial number of training process of the dataset was required (Li et al. 2021).

The steps to apply the ANN-MLP are described as follows:

(a) Data Training Process

(i) Phase 1: Under the quick tab the datasets were randomly selected by the software prior to the analysis into three subsets: (i) training subset (70%) used for training the model, (ii) testing subset (15%) which is necessary for monitoring or improving the network quality from over-fitting or over-training of the generated model, and (iii) validation subset (15%) representing the validation purposes of the ANN generated model, but not creating the ANN model network. In total, the seed for sampling (seed for network initialization) was automatically set as 1000 (seed for sampling can be manually changed depending on the subset required), 50 networks were tested, whereas 10 retained. In the learning phase, the optimization on the neurons in the hidden layers and the connection weight between neurons was performed (Zupan and Gasteiger 1999). The cross-entropy was set as the cross-entropy error function, and the number of networks to be performed was also specified. However, the cross-entropy error function is applicable only for classification problem analysis, while the sum square error function only applicable for regression problems analysis. When using cross-entropy as the error function, thus the MLP output activation function was automatically restricted to softmax. This restriction as to ensure the accuracy and high precision obtained for the class membership probabilities (classification). The activation function of the hidden units was selected such as tanh (uses the hyperbolic tangent function and the output within the range of $(-1, +1)$).

(ii) Phase 2: The multilayer perceptron (MLP) as the network type was selected using the training algorithm BFGS with 200 iteration cycles with the stopping condition of $1.0 \times 10{-}7$ as change in error, while window set at 20. Normal randomization was set as the network randomization with variance/ max set as 0.1.

(iii) Phase 3: Further, made the selection on the weight decay tab to allow the weight decay regularization specified. The application proceeded with the use of weight decay for hidden layer and output layer with the values of 0.01 and 0.001, respectively. The selection made of weight decay enables the weight decay regularization to apply to the input-hidden layer weight and hidden-output layer weight. In particular, the regularization functions are for the network error functions modification, of which discouraging (penalizing) the large weight values where the additional term of E_w was added in the form of the following equation (Eq. 5.5):

$$E = E_{\text{sos}} + E_w$$

$$E_w = \frac{\infty}{2} w^T w \tag{5.5}$$

where α denotes as the weight decay constant, w^T and w both are the network weights. The greater value of α, the more weight is penalized as to reduce the over-fitting problems.

(iv) Then, the initialization tab was selected with the seed initialization of dataset as 1000 and finally chose the real time training graph tab. The real-time training graph was restricted to the train error and test error. Finally, run the training process.

The network architecture of ANN-MLP, created for oil type classification and regression from the oil spill compounds data matrix consisting of 94 petroleum hydrocarbon compounds from 47 oil spill samples as the neurons of the input layer, two hidden layers (1–24 neurons) and one output layer. The ANN-MLP feed-forward propagation offered the closely integrated among the artificial neurons or nodes in each hidden layer and the layer of output (prediction) with the input layer (corresponding layer) (Cajka et al. 2010). The interconnection between the neurons of ANN built by the weight or numerical weight passes the magnitude signal among them for the classification and regression.

In MLP network architecture rule, the signals are transferred between neurons from the previous layer via the hyperbolic tangent function of the transfer function (i.e., mathematical function). The reason was to substantially acquire the accurate classification or regression (prediction) of the oil spill data output, but subject on the common basis that the learning or training process continues until the synaptic weights and bias reach the acceptable level (Cajka et al. 2010). The learning process continues once the individual training process complete. On the other word, the training process stops when the error function stops improving. The threshold value in each node or neuron is called bias. The activation function of the neuron functions has strong influence in ANN-MLP, transforming the artificial neurons (each unit) of the input to tanh function introduces nonlinearity classification in ANN model), while the output activation function sets to identity for regression. The mathematical equation is expressed as below (Eq. 5.6) (Houhou and Bocklitz 2021):

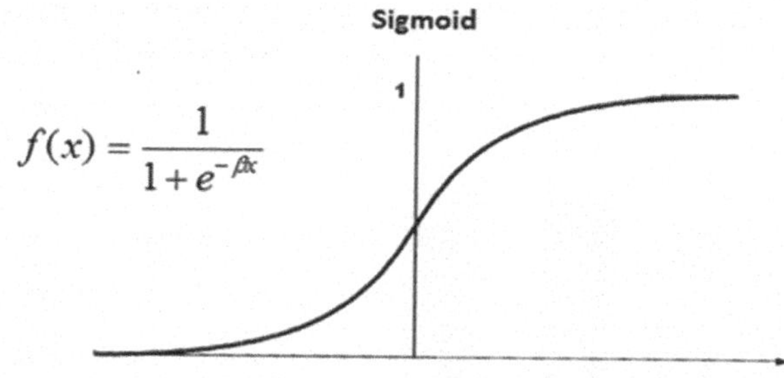

Fig. 5.2 An illustration of the S-curve for the output activation function of tanh

$$f(x) = \frac{e^a - e^{-a}}{e^a + e^{-a}} \tag{5.6}$$

The following equation of ANN expressed the performance of the datasets classification;

$$\mu_i = \sum_j^N = 1^{\text{Wijxj}} \tag{5.7}$$

f(x) or f(ui) is the neuron output of *i*, xj is the neuron input of *j*, and yi is the weighted sum output of the i inputs of N neurons of hidden or proceeding layer that connected with the i neuron. Figure 5.2 is the illustration of the S-curve for the output activation function of tanh, (hyperbolic tangent activation function).

To give a further profound understanding, while the network was in use, the threshold value of each input variable of the oil compounds is correctly placed in each artificial input node. Then, once established, the nodes in proceeding layers and output layer were progressively built. In addition to the network, the threshold value of each node dealt with the activation value that each neuron individually calculates the activation value by adjusting the weighted sum output of yi of the i inputs of N neurons of proceeding layer, which corresponds with the reduction of threshold value (bias). The adjustment of the weighted sum and threshold value (bias) was to minimize the error in the prediction of the actual output. Subsequently, the activation value in each node was transferred to the output node (yi) in the output layer through the transfer function (Fig. 5.3).

Moreover, the successive training of ANN-MLP provides huge influence in terms of relationship between the input and output layers in learning phase. The output layer was built once all the networks were used/trained, and the output (yi) of the output layer serves the entire output of the entire network.

The testing set of data in ANN-MLP technique was paramount important as to improve the over-fitting or over-training of the network (Wu and Feng 2018). Also, it was used to determine whether the ANN model the one workable and to check whether the created model worked well on the data that previously never been trained.

THE PREDICTION OF OIL SPILL TYPES BY PCA & ARTIFICIAL NEURAL NETWORK

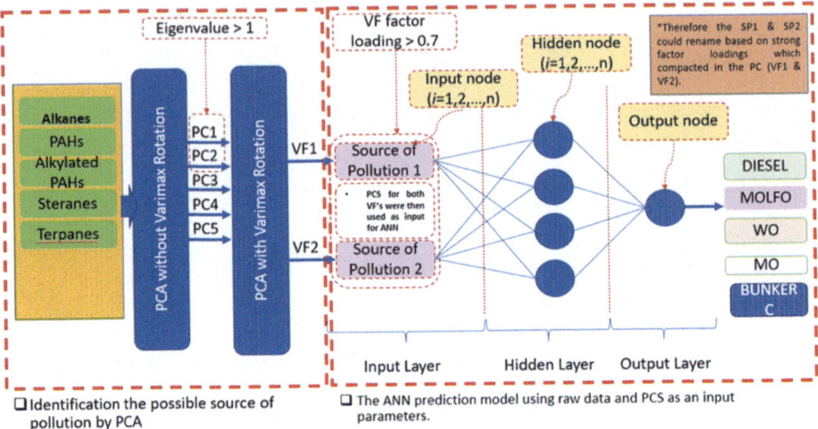

Fig. 5.3 Artificial Neural Network–multilayer perceptron (ANN-MLP) architecture model for oil type classification

References

Cajka T, Riddellova K, Klimankova E, Cerna M, Pudil F, Hajslova J (2010) Traceability of olive oil based on volatiles pattern and multivariate analysis. Food Chem 121:282–289. https://doi.org/10.1016/j.foodchem.2009.12.011

Houhou R, Bocklitz T (2021) Trends in artificial intelligence, machine learning, and chemometrics applied to chemical data. Anal Sci Adv 2:128–141. https://doi.org/10.1002/ansa.202000162

Ismail S, Reza MF, Jusoh MH, Wan Awang WS, Juahir H (2023) Psychospiritual healing from al-Quran: internal aesthetic factor of Quranic sound and its effects in activating greater brain regions. Malays J Fundam Appl Sci 19:583–606. https://doi.org/10.11113/mjfas.v19n4.2969

Schmidhuber J (2015) On learning to think: algorithmic information theory for novel combinations of reinforcement learning controllers and recurrent neural world models. Tech Rep, arXiv:1511.09249v1 [cs.AI]. https://arxiv.org/abs/1511.09249

Li LN, Liu XF, Yang F, Xu WM, Wang JY, Shu R (2021) A review of artificial neural network based chemometrics applied in laser-induced breakdown spectroscopy analysis. Spectrochim Acta B at Spectrosc 180:106183. https://doi.org/10.1016/j.sab.2021.106183

Wu YC, Feng JW (2018) Development and application of artificial neural network. Wirel Pers Commun 102:1645–1656. https://doi.org/10.1007/s11277-017-5224-x

Zupan J, Gasteiger J (1999) Neural networks in chemistry and drug design. John Wiley & Sons, New York

Chapter 6
Bridging Machine Learning and Oil Spill Data

Abstract This chapter described the details about the implementation of machine learning in oil spill. Besides that, this chapter explores the application of Support Vector Machines (SVMs) as an advanced machine learning technique, focusing in oil spill fingerprinting, and a vital component on environmental forensics. The aim of this chapter is to accurately classify and predict the origin of oil spills by analyzing complex chemical compositions and environmental parameters. The key advantages of the SVM approach are high predictive accuracy, fast computational performance, and also reliable separation of complex chemical profiles. Through kernel-based mapping and margin maximization, the SVM model demonstrates its potential in identifying oil spill sources, supporting legal accountability, and enhancing mitigation planning. The chapter concludes that SVMs provide a scientifically sound and computationally efficient methodology for tackling real-world environmental classification challenges.

Keywords Support vector machines · Machine learning · Kernel-based mapping · Margin maximization · Predictive

6.1 Introduction

Similar to Artificial Neural Networks, Support Vector Machines (SVMs) are another clustering-based, unsupervised methodology for distinguishing the nonlinear boundaries of the dominating population from the remaining data. This is a machine learning technique and the most effective clustering algorithm for resolving pattern recognition issues in practical applications. This method was initially presented as the most effective and well-known clustering algorithm. Oil spill fingerprinting requires a crucial procedure in environmental forensics. Legal accountability, planning for mitigation, and comprehending the ecological effects of spills all depend on this. The main purpose of using this technique enabled the optimal hyperplane of classification (maximum margin hyperplane) and regression of huge and complex datasets involving chemical compositions and environmental characteristics. The use of the

A. Ismail and H. Juahir, *Advanced Chemometrics*,
SpringerBriefs in Environmental Science, https://doi.org/10.1007/978-3-032-10742-8_6

SVM model as an unsupervised machine learning tool has made it possible to improve the oil recovery process with a high degree of predictive accuracy achievable within a short execution period (fast computation time).

SVM responsible for separating the oil type based on chemical compound properties in optimally margin hyperplane. In order to separate or classify the oil spilled variables into one or two classes or categories with comparable homogeneity, the SVM application in this book was used to convert the input features of the oil spill compounds (unseen variables) into a nonlinearity separable high-dimensional hyperplane. A real classification and regression (prediction) of the real-valued output of yi with the real-input value of xi was the rationale behind the application of the SVM.

This technique was to predict the type of oil spills from the distinctive origin. The oil spill dataset was analyzed for clustering and pattern recognition according to its intrinsic chemical properties. It derived meaningful complex dataset for precision and reliability using the multiple hyperplanes. There were four steps of the SVM model nonlinear mapping approach used in this book, which are clearly described below.

6.2 Implementing Classification and Regression Models

6.2.1 Data Pre-treatment

From the laboratory analysis, 94 oil spill compounds were yielded used as the input data from 47 distinctive oil spill samples. The whole dataset or input dataset was undergone normalization as it's are all numerical to the dataset's limited region of yi € (+1, −1) as to avoid truncation errors (Adib et al. 2013).

$$\mathrm{Xn}_i = 2\frac{X_i - X_{\min}}{X_{\max} - X_{\min}} - 1 \tag{6.1}$$

X_i is the actual input/output oil spilled data, while (xn_i) is the collection of scaled input or output oil spilled data. The observation dataset's smallest value is denoted by $X_{\min}$, while its maximum value is denoted by Xmax. The output model of yi was also standardized since the input dataset was normalized using Eq. 6.1. Before the data was pre-treatment or processed, the input dataset of oil spills was mapped into the normalized feature space to get the output variables. In data analytics, Support Vector Machines (SVMs) can be roughly divided into three subcategories according to the kind of learning and challenge. Upon completion of the pre-treatment, the data was utilized for three categories of analyses, namely the classification, regression Support Vector Clustering. The mathematical formulation plays a pivotal role as succinctly described below.

- These labels and input vectors are used by the model to determine the best hyperplane with the largest margin between classes.

- If the input data is nonlinearly separable, it may be mapped onto a higher-dimensional space using the kernel trick (such as the RBF kernel). The approaches of the three categories are briefly explained as below the following.

6.2.1.1 C-SVM Classification

The classification of SVM is to classify data into discrete classes by determining the best hyperplane to divide the data points into distinct classes by the greatest amount. It involves with three subsets—training, optimization, and testing—that are randomly selected from the normalized dataset (Adib et al. 2013). Interestingly, 80% of the data were used for training subsets, 10% were used for optimization, and 10% were utilized for testing, respectively. Kernel-RBF for nonlinearity: this kernel method is applicable to both classification and regression of oil spill dataset. The fundamental mechanism of Support Vector Machines' (SVM) approach to oil spill data classification is explained in the following data training process, optimization, and regression of mathematical formulation for an SVM model.

6.2.1.2 Data Training Process

(i) Phase 1: The training dataset was provided using the formula, where N is the number of oil-spill samples. Each of the N samples that make up the dataset represents oil spill's types.
(ii) Phase 2: Input Vector ($X \in \mathbb{R}^n$): X € Rn is the input vector of n input features and labels for Classes (yi $\in \{+1, -1\}$). The yi € (+1, −1) is a finite value with class labels or a bounded region. The complexity of the data is captured by the notation $X \in \mathbb{R}^n$, which shows that each input vector X exists in an n-dimensional real-valued space.

The function of yi € (+1, −1), but yi € 1,2,3,4,5,6,7,8,9,0) in binary classification (Rodriguez-Galiano et al. 2015). For every input vector X, yi stands for the class label. Therefore, the function of + 1 and −1 in the classification of oil spills is described as below:

- + 1: Indicates that an oil spill has been positively identified.
- − 1: Denotes a spill that isn't oil, like water, algae, or other mimics.

SVM relies heavily on this binary classification method, where the model learns to draw a hyperplane that best separates these two classes. The complexity of the data is captured by the notation $X \in \mathbb{R}^n$, which shows that each input vector X exists in an n-dimensional real-valued space.

Three important parameters as follows were established before the dataset was trained in order to maximize the SVM model:

- The function's smoothness is controlled by γ (Gamma), which also helps to lower fitting mistakes.

- The precision level, or permissible departure from the true value, is defined by ε (Epsilon).
- The SVM model's overall efficacy is influenced by σ2 (sigma squared). (Adib et al. 2013). The provided formulas, which include non-negative slack variables $\xi n * \xi$ n ∗, and $\epsilon^{\cdot}$, provide an upper and lower constraint on the difference between y (n) and f (Xn). This frequently occurs in support vector regression, regression models, and optimization problems (Eq. 6.2).

$$y(n) - f(\text{Xn}) \leq \xi n + \varepsilon$$

$$f(\text{Xn}) - y(n) \leq \xi_n^* + \varepsilon$$

$$\varepsilon, \xi_n, \xi_n^* \geq 0, \forall n. \tag{6.2}$$

Slack variables ξ, which take into account both positive and negative classes, quantify the degree to which data points depart from the ideal categorization border. How the model responds to misclassification errors is referred to as sensitivity. The best values of the main parameters (γ, ε, σ2), where γ stood for the estimated function smoothness and fitting error minimization, ε for the accuracy threshold, and σ2 for the efficiency of the SVM model, were used to define the separating hyperplane (training the dataset). The slack variable (ξ) is used to assess how successfully the SVM model classifies data points while taking into account both positive and negative classes. Owing to the classification boundary is somewhat flexible, it aids in managing misclassified points. The model's estimated prediction is represented by the real-valued output (yi), as cited by (García-Nieto et al. 2018).

In simple terms:

As a collection of real-valued values, xk (x_1, x_2, x_3...xn) represents the input features.

The direction of the dividing border, or hyperplane, in a high-dimensional space is defined by the vector w.

The bias term b causes the hyperplane to move away from the origin.

To make sure that no data points fall between two hyperplanes, a margin is created for linearly separable data. (Nieto et al. 2018).

$$\text{Wxk} - b = 1 \tag{6.3}$$

$$\text{Wxk} - b = -1 \tag{6.4}$$

6.3 Optimizing the Kernel

For handling complex, nonlinear data in oil spill detection, the Radial Basis Function (RBF) kernel is frequently chosen. There are two setting up parameters commonly applied in kernel optimization.

Gamma (γ): Regulates the impact of isolated data points. Higher numbers highlight more intimate points, and lower values indicate a wider influence.

Regularization parameter, or C: Achieving a low error on training data while keeping a straightforward model to prevent overfitting is balanced by the C.

6.3.1 Regression SVM (Support Vector Regression—SVR)

The regression SVM is also known as Support Vector Regression also known as epsilon-SVM regression (Liu et al. 2022). It determines the optimal hyperplane within a tolerance margin to the training data (6.5), which predicts continuous values instead of categorical labels. It predicts numbers (such as oil spill's chemical compounds) by finding the best-fit line (or curve) within a predetermined margin of error, rather than categorizing data into "oil spill types" or "not oil spill types" categories.

$$\frac{1}{2}W^T W + C\sum_{i=1}^{n}\xi_i + C\sum_{i=1}^{n}*\xi_i \tag{6.5}$$

This regression technique was utilized to categorize and forecast results while permitting some error tolerance in order to manage nonlinear data. The model was trained by gradually optimizing the error function. To increase accuracy for chemical analysis of oil spills, two SVM models (C-SVM and Support Vector Regression—SVR) were chosen. The equation for C-SVM classification is trivially expressed as follows once the error function minimization for this C-SVM was resolved (García-Nieto et al. 2018).

Kernel-RBF (Radial Basis Functions) provided the best marginal hyperplanes. By converting low-dimensional input space into higher-dimensional space, a kernel is a function that turns non-separable problems into separable ones. When implementing kernel-based learning approaches, the most popular and reliable method is the case-wise approach. Additionally, the center of the subsets with two tuning parameters (γ, α) was classified using a specific nonlinear classification function in the Kernel-RBF approach. The kernel parameter is denoted by α, while the regularization parameter is denoted by γ. The statistical inference analysis is genuinely centered around these functions. For nonlinearity classification and regression, the training cost constant capacity, C, is set at 10 and the constant γ value is set at 0.185. When a regression is nonlinearly separable in the input space, the kernel trick is utilized to solve the problem.

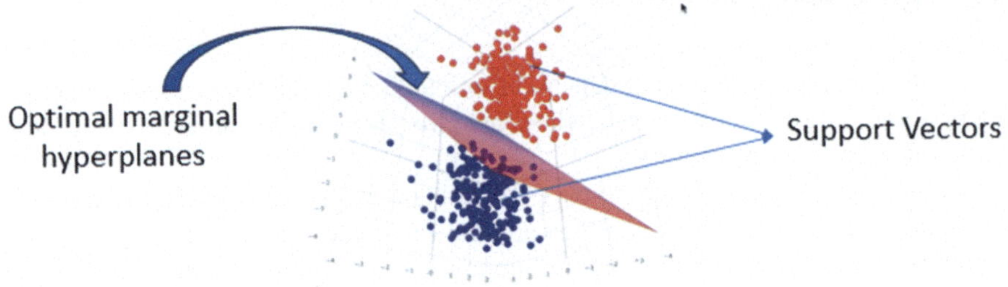

Fig. 6.1 Support vector machine hyperplane. *Source* google/ November 20, 2024/11.23 am (Malaysia Time))

A set of functions that formed the arbitrary basis as the decision boundary for oil spill compounds were provided by the hidden layer in Kernel-RBF. Following that, these functions were transformed into the input layer of oil spill compounds (SAS141A, SAS141B, BAS303A, BAS304B,…Xmo) or input vectors. These input vectors were then expanded or spanned through the hidden layer of kernel inner product (support vectors (xi)), to which the linear output layer mapped the input vectors (training vectors) to optimal discrimination as the output (y) (Fig. 6.1). In the output feature space of an Artificial Neural Network (ANN), the kernel inner product offered the best hyperplane design or comparable function of the neuron network. The appropriate weights and support vectors (independent variables) were calculated correctly during the training phase, and only those support vectors produced a nonzero α coefficient.

6.3.2 *The Importance of Support Vector Machine (SVM)-Based Predictive Model Validation*

A Support Vector Machine (SVM) model's accuracy, dependability, and capacity for generalization are all dependent on predictive model validation (Zhang and O'Donnell 2020). In order to avoid problems like overfitting (when the model remembers training data but fails on new inputs) or underfitting (when the model is too simplistic to capture patterns), validation helps evaluate how well the trained SVM model works on unseen data. To ensure an ideal balance between bias and variance, methods like k-fold cross-validation and grid search optimization aid in fine-tuning important parameters like the kernel type, regularization parameter (C), and kernel coefficient (γ). Validation assesses the model's capacity to accurately distinguish between classes in classification, whereas regression (SVR) assesses the accuracy of continuous value predictions. In regression (SVR), validation assesses the accuracy of continuous value predictions, whereas in classification, it gauges the model's capacity to accurately distinguish between classes. Accurate decision-making in a variety of

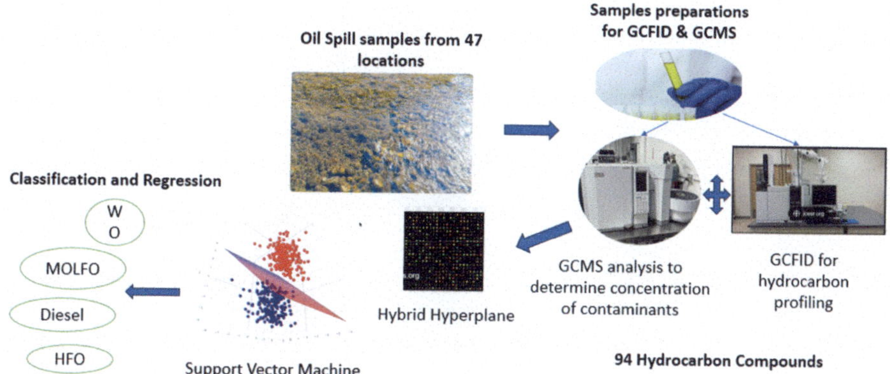

Fig. 6.2 Process of making classification and regression by hyperplane support vector machine (SVM)

applications, including financial forecasting, medical diagnosis, and environmental monitoring, is made possible by rigorous validation of SVM models.

Using a tenfold cross-validation procedure carried out by the cross-validation algorithm, the cross-validation method was chosen for predictive model validation. This was done by evaluating the optimal value of various variables to determine the Mean Error Square (MES), coefficient determination (R2) of the fitness factor, and correlation coefficient.

The entire process of Support Vector Machines (SVMs) dividing data for both classification and regression in oil spill fingerprinting tasks using a hyperplane involving the complex relationships in numerical data is illustrated as in Fig. 6.2.

References

Adib H, Sharifi F, Mehranbod N, Kazerooni NM, Koolivand M (2013) Support vector machine based modeling of an industrial natural gas sweetening plant. J Nat Gas Sci Eng 14:121–131. https://doi.org/10.1016/j.jngse.2013.06.004

García-Nieto PJ, García-Gonzalo E, Alonso Fernández JR, Díaz Muñiz C (2018) Predictive modelling of eutrophication in the Pozón de la Dolores lake (Northern Spain) by using an evolutionary support vector machines approach. J Math Biol 76:817–840. https://doi.org/10.1007/s00285-017-1161-2

Liu T, Jin L, Zhong C, Xue F (2022) Study of thermal sensation prediction model based on support vector classification (SVC) algorithm with data preprocessing. J Build Eng 48:103919. https://doi.org/10.1016/j.jobe.2021.103919

Rodriguez-Galiano V, Sanchez-Castillo M, Chica-Olmo M, Chica-Rivas M (2015) Machine learning predictive models for mineral prospectivity: an evaluation of neural networks, random forest, regression trees and support vector machines. Ore Geol Rev 71:804–818

Zhang F, O'Donnell LJ (2020) Support vector regression. In: Elsevier (ed) Reference module in neuroscience and biobehavioral psychology. Elsevier, Amsterdam. https://doi.org/10.1016/B978-0-12-815739-8.00007-9

The manufacturer's authorised representative in the EU is Springer Nature Customer Service Centre GmbH, Europaplatz 3, 69115 Heidelberg, Germany. If you have any concerns regarding our products, please contact ProductSafety@springernature.com

Printed and bound by CPI Group (UK) Ltd, Croydon, CR0 4YY
07/07/2026
02160927-0002